T0135214

INTRODUCTION TO PROTEOMICS

INTRODUCTION TO PROTEOMICS

Principles and Applications

Nawin Mishra

A JOHN WILEY & SONS, INC., PUBLICATION

Cover: Proteomics of Metamorphosis in Insect - A Computer Projection.

Published by John Wiley & Sons, Inc., Hoboken, New Jersey
Published simultaneously in Canada

For general information on our other products and services or for technical support, please contact our Customer Care Department within the United States at (800) 762-2974, outside the United States at (317) 572-3993 or fax (317) 572-4002.

Wiley also publishes its books in a variety of electronic formats. Some content that appears in print may not be available in electronic formats. For more information about Wiley products, visit our web site at www.wiley.com.

Library of Congress Cataloging-in-Publication Data:

Mishra, N. C. (Nawin C.)
Introduction to proteomics : principles and applications / Nawin Mishra.
p. ; cm.—(Methods of biochemical analysis ; 52)
Includes bibliographical references and index.
ISBN 978-0-471-75402-2 (paperback)
1. Proteomics—Textbooks. I. Title. II. Series: Methods of biochemical analysis ; v. 52.
[DNLM: 1. Proteomics. 2. Proteome—analysis. W1 ME9617 v. 52 2010 / QU 58.5 M678i 2010]
QP551.M475 2010
572′.6—dc22

2009049260

Printed in the United States of America

10 9 8 7 6 5 4 3 2

This book is dedicated to the memory of
Professer E. L. Tatum
and my parents, the mentors in my life,
and to Purnima and Prakash.

CONTENTS

FOREWORD

Proteomics provides a better understanding of cells by elucidating the structure, function, and interactions of proteins. The one gene–one enzyme concept of Beadle and Tatum provided an important tool necessary for the analysis of proteins by creating a mutant protein and then comparing its properties with that of the wild-type protein. This method of Beadle and Tatum and the method of Edman degradation have become standard tools for deciphering the structure and function of proteins until the coming of genomics and the high-throughput methods of mass spectrometry and bioinformatics. In this context, the book on *Introduction to Proteomics* by Nawin Mishra, who was an associate of Tatum at a time when the structure and function of proteins were being elucidated in laboratories around the world, is important. This book deals with all the basic and medical aspects of proteomics, including personalized medicine. This book could serve as a valuable reference for all those interested in proteomics.

GÜNTER BLOBEL
Laboratory of Cell Biology
The Howard Hughes Medical Institute
The Rockefeller University
1230 York Avenue
New York, NY 10065-6399

PREFACE

Proteomics is the study of all the proteins of a cell or an organism. It is the newly developed science for the study of proteins. It attempts to define the proteome, which is the entire protein content of an organism encoded by its genome; hence, the word is derived from protein and genome. Proteomics aims at describing the structure and function of the proteins of a cell at a large scale. This enables us to understand the structure and function of a cell and finally that of an organism. The science of proteomics has obvious applications to medicine through identification of proteins as marker(s) of a disease (i.e., diagnostics) or as targets of new drugs or as therapeutics (i.e., drugs) as well. Proteomics provides new tools for the understanding of proteins, which are the workhorse molecules of a cell that control all its biophysical and biochemical attributes. The one gene–one enzyme concept of Beadle and Tatum (1941) provided a unique tool for the study of proteins; this approach is being used every day, even to this date. Proteomics based on high-throughput technologies added a new dimension to the approach initiated by Beadle and Tatum. This book, therefore, examines proteomics beyond the one gene–one enzyme concept.

My research interest in genetics and the biochemistry of proteins goes back to the mid-1960s, when I began my association with the late Nobel Laureate Professor Edward L. Tatum at the Rockefeller University as a postdoctoral fellow supported by the Jane Coffin Childs funds for Medical Research. Beadle and Tatum together formulated the one-gene–one enzyme concept in 1941. George Beadle, Edward L. Tatum, and Joshua Lederberg shared the 1958 Nobel Prize in Physiology and Medicine for their respective contributions to the development of the one-gene–one-enzyme concept in Neurospora and recombination in bacteria; Lederberg later became president of Rockefeller University. This theory of Beadle and Tatum established the conceptual scheme for the control of the structure and function of a protein by a gene.

At Rockefeller University, the laboratories of William Stein and Stanford Moore and that of Robert Bruce Merrifield were situated close to Tatum's laboratory. In their laboratories, the first large protein was sequenced and chemically synthesized. I remember having several discussions with these scientists about the structure and function of proteins. William Stein, Stanford Moore, and Gerry Edelman, all of whom were from Rockefeller University, and Christian Anfinsen of the National Institutes of Health (NIH) became Nobel Laureates in 1972. Later, Bruce Merrifield in 1984 and Günter Blobel in 1999, also from Rockefeller University, received Nobel Prizes, all of them for their contributions to protein chemistry, including the structure, function, synthesis, and intracellular transport of proteins. The goal of Stein and Moore at that time was to sequence more than 1000 proteins by the end of the 20th century. This goal was realized much faster with the science of genomics and with the application of mass spectrometry and other high-throughput technologies.

At Rockefeller University, I also had the opportunity to know Professor Frank H. Field, director of the mass spectrometry laboratory. Earlier, Dr. Field, in collaboration with Joe Franklin, had developed the first ionization technique for mass spectrometry. Dr. Field was helping Professor Tatum with the identification of chemical(s) emitted into the gas phase by a slow-growing morphological mutant of Neurospora. An exposure of this gaseous emission to the wild-type strain made it grow slowly like the mutant. This chemical, however, remained elusive to identification by mass spectrometry.

Soon after my arrival at Rockefeller University, I remember having a discussion with the Professor Victor Najjar on the one-gene–one-enzyme theory. Dr. Najjar, then a Professor at the Vanderbilt University and an editor of *Methods in Enzymology*, was visiting Rockefeller University on a sabbatical leave. During a discussion of my work with him, he became somewhat concerned after learning about the possible role of two genes in the control of an enzyme, phosphoglucomutase, involved in the morphogenesis of a fungus Neurospora as my work indicated at that time. I believe this was perhaps because of his unfamiliarity with the literature in genetics and particularly that of the role of suppressor genes in controlling the structure of a protein encoded by another gene. He, therefore, thought that my findings were in contradiction to the original idea of the one-gene–one-enzyme hypothesis. However, I convinced Dr. Najjar that such findings make a difference only in semantics and not in the conceptual scheme of the original one-gene–one-enzyme theory. I pointed out to him that these exceptions only strengthen the original one-gene–one-enzyme concept, just as certain observations such as the partial dominance, co-dominance, and epistasis, which on the surface seem to be in conflict with Mendelian rules of inheritance, actually lend support to the original ideas implicit in the rules of inheritance by Mendel.

Later that day, I discussed with Professor Tatum the exchange on the one-gene–one-enzyme theory during my conversation with Dr. Najjar. During our conversation, Professor Tatum immediately pointed out that the one-gene–one-enzyme hypothesis has already been modified to a one-cistron (gene)–one-polypeptide hypothesis: However, I was aware of this concept and told professor Tatum that I had already pointed out this modification to Professor Najjar. Professor Tatum also expressed that he expected additional modification to this theory because of the looming complexity of our genetic material as was being revealed by the nucleic acid hybridization experiments. He expressed to me that it was indeed a matter of semantics and that so long we understood what we were talking about, we lived with the limits of the conceptual scheme of the one-gene–one-enzyme hypothesis. Almost a decade later, Phillip Sharp from the Massachusetts Institute of Technology (MIT) revealed the split nature of the gene and received the Nobel Prize in 1990 for his work. Furthermore, the study of the structure of the immunoglobulin gene(s), which brought the Nobel Prize to Tonegawa, also from MIT in 1987, presented an extreme view of an exception to the one-gene–one-enzyme hyothesis. However, these findings affirmed the expectations of Professor Tatum that the one-gene–one-enzyme theory would be modified in view of the complexity of our genetic material. Despite the changes to this theory, it is important to note that almost all genes in prokaryotes and more than 50% of genes in higher eukaryotes obey the dictum of one-gene–one-enzyme theory. This theory still provides the basis for creation of mutants and knockouts crucial for the study of a protein structure and function and its role in controlling the phenotype of the organism. This theory is also the basis for the gene therapy approach for the treatment of human diseases.

I remember the events and the manner in which the field of protein chemistry progressed and then was later ignored with the coming of the genome projects and the science of genomics; it was finally revived and blossomed into the science of proteomics. The coming of genomics and the subsequent development of proteomics have completely changed our view regarding the philosophy of science and how we understand biology. Before genomics, we had a reductionist view of science, and the biology of an organism was thought to be understood in terms of the molecules only. We also used to do one thing at a time when deciphering one molecule after another. Now, we are trying to understand all things at the same time because of our ability for high-throughput analyses; we are no longer reductionists, rather we are holists trying to understand the biology in terms of the interactions of a large number of molecules at once. The science of proteomics has thus ushered in the coming of a new branch of science called *systems biology* to obtain the ultimate understanding of an organism within a particular environment. An understanding of the environment is important because it can bring about changes in the structure and function of genes and gene products.

I write this book on the science of proteomics with the goal of bringing out its conceptual development starting from one-gene–one-enzyme theory and leading to its instrumentation-based methodologies and applications in medicine and biotechnology and the fact that life is sustained by the interactions of proteins. I take special effort in describing the nature and operation of these complex instrumentations involved in proteomics in a language readily understandable to students with an exclusive background in biology. I also provide an emphasis on biological methods in elucidating certain aspects of proteomics, which has been ignored in earlier treatises on the subject of proteomics. This book is written in a manner comprehensible to emerging scientists, including undergraduate and graduate students as well as postdoctoral trainees.

The book is organized into seven chapters, and many references, although some included at the end of the chapters, are not cited in the text to allow for the smooth flow of main concepts and easy reading of the subject matter. I hope that my efforts are successful.

I believe no such text that particularly addresses the needs of the biologist exists at this time. In this book, an attempt is made to give a biologist's view of the subject to non–biologists equally well, particularly bringing to their attention how biologists approached certain problems—for example, protein–protein interactions in the absence of advanced technologies such as bioinformatics. I also believe that this text is a contribution to this emerging branch of science of proteomics and to systems biology, and of course to scientists in these branches of science, leading to the appreciation of the developments in proteomics beyond the one-gene–one-enzyme concept of Beadle and Tatum that provided the conceptual scheme and the tool for understanding proteins in the living system.

This book is being published on the occasion of the 52nd anniversary of the awarding of the Nobel Prize to Beadle and Tatum in 1958 to reflect the progress made in the understanding of proteins, which was started by the conceptualization of the one-gene–one-enzyme hypothesis that provided the tool for analysis of proteins.

I would like to thank many colleagues for their help with this work. I would like to thank Professors Steve Threlkeld and J.J. Miller, both of McMaster University, for my fueling initial interest in genetics and Professor Stuart Brody of the University of California, San Diego, (formerly at the Rockefeller University) for my introduction to enzymology. In addition, I am grateful to Professor Philip Hanawalt of Stanford University and Professor Stuart Linn of the University of California, Berkeley for their support of my continued interest in the genetical biochemistry of proteins. I would also like to thank Professor David Reisman at the University of South Carolina for reading the manuscript in its entirety and for his many helpful comments. I am also thankful to Professors Michael Felder and Sanjib Mishra both at the University of South Carolina, Professor Narsingh Deo of the University of Central Florida, Professor David Gangemi of Clemson University, Professor Alexandru Almasan of the Cleveland Clinic, Dr. Narendra Singh of the U.S.C. Medical School, Professor R.P. Jha of Patna University, Professor K.M. Marimuthu of the Post Graduate School at Madras University, Professor Ramesh Maheshwari of the Indian Institute of Science, Prashant Jha and Dr. Kanchan Kumari for their support of my endeavors and to Dr. Richard Vogt of the University of South Carolina for help with the cover picture.

This work would not have been possible without the encouragement and show of infinite patience from Dr. Darla Henderson of John Wiley and Sons, particularly during periods of multiple personal challenges. I also thank Anita Lekhwani, the Senior Acquisition Editor of John Wiley and Sons, for her immense interest in this work and for her enthusiastic support and assistance that eased the submission of this manuscript and made its publication possible. I am also thankful to Christine Moore, Rebekah Amos, Sheree Van Vreede, and Kellsee Chu of John Wiley & Sons for assistance with the manuscript that helped its timely publication. I am grateful to Dr. Kevin H. Lee of the University of Delaware for the two-dimensional gel picture, Darryl Leza of NHGRI, NIH, for the protein structure picture, and to John Alam, Clint Cook and Michelle J. Bridge of the Dept. of Biological Sciences at the University of South Carolina for the diagrams and for their assistance in preparation of the manuscript.

Finally, I thank my wife, Purnima, and our son, Prakash, for their continuous support and interest in this work. I dedicate this work to Purnima and Prakash and above all to the memory of the mentors in my life, my parents and Professor E.L. Tatum. I am solely responsible for any and all errors that may be found in this book.

ABOUT THE AUTHOR

Nawin Mishra received his PhD. in genetics from McMaster University in 1967. His postdoctoral training was with the late Nobel Laureate Professor E. L. Tatum at Rockefeller University, supported by a postdoctoral fellowship from the Jane Coffin Childs Memorial Fund for Medical Research at Yale University. In 1973, he joined the molecular biology faculty of the University of South Carolina as an associate professor; he remained there as Distinguished Professor of Genetics until 2006. Currently, Dr. Mishra is still with the University of South Carolina as Emeritus Distinguished Professor of genetics. Dr. Mishra was a visiting professor at the Max Planck Institute of Molecular Biology in Heidelberg, Germany, in 1980 and at the Greenwood Genetics Center in 2004. He initiated the gene-transfer experiments in fungi while he was a member of the laboratory of Dr. E. L. Tatum at Rockefeller University (1967–1973). He has investigated various aspects of gene transfer, the organization of mDNA, and the biochemical genetic characterization of proteins in carbohydrate and DNA metabolism.

Dr. Mishra has been invited to present his work in Australia, Europe, Russia, China, Japan, Thailand, and India. He served as a Scientific Consultant to the Food and Agriculture Organization (FAO) of the United Nations in 1990 and in 1993. He also served as Chairman of the Program Committee of the Genetics Society of America and as a member of the review panel of the Human Genome Project of the U.S. Department of Energy. He has served as a fellow of the American Association for the Advancement of Science since his election to this organization in 1986 for his original contributions to the study of gene transfer in fungi. Dr. Mishra has organized the Genetics

Society of America annual meeting in 1978 and the First Fungal Genetics Congress in 1986; he has also written a book that was first published by CRC Press in 1995, and whose expanded version was published by John Wiley & Sons in 2002.

CHAPTER 1

HISTORICAL PERSPECTIVES

Biology becomes much more understandable in light of genetics (Ayala and Kiger 1984). This is true even more so in the case of the theory of evolution proposed by Darwin (1859). It seems the theory of evolution would have been placed on a solid foundation from the start if Darwin would have been aware of the Mendelian rules of inheritance. There is some indication that a copy of Mendel's publication was received by Darwin, which remained unopened during his lifetime. It is believed that this caused Darwin's failure to provide a firm basis on which selection works during the process of evolution.

Genetics has had several major breakthroughs during its development that have made biology a well-established discipline of science. Some of these break throughs are discussed here. The first major discovery was the rules of inheritance by Mendel (1866). This provided the particulate nature of inheritance and established the presence of genes, which control phenotypes. It also provided genes as the ultimate basis for propelling the process of evolution of organisms and integrated the different branches of the science of biology. In addition, Mendelian genetics transformed biology from a science based exclusively on observations to an experimental science where certain ideas could be tested by performing experiments.

The second major breakthrough was discovered by Beadle and Tatum (1941) with their conceptual one-gene–one-enzyme hypothesis. This proved the biochemical basis for the mechanism of gene action and integrated

Introduction to Proteomics: Principles and Applications, By Nawin C. Mishra

chemistry into biology. It provided the tool for analyzing metabolic pathways and several complex systems, including the nervous system. It also provided the understanding of the genetic basis of diseases and their possible cures by chemical manipulations and ultimately by gene therapy.

The discovery of the structure of DNA by Watson and Crick (1953) marked the third major breakthrough in biology. The discovery of the Watson–Crick DNA structure was aptly meaningful in view of the findings of DNA as the chemical basis of inheritance (Avery et al. 1944, Hershey and Chase 1952). The Watson–Crick structure of DNA provided the molecular basis for the understanding of the mechanisms of the storage and transmission of genetic information and possible changes (mutations) therein. Mutation provided the source of variations that could be selected for during the process of Darwinian evolution. Thus, the DNA structure created by Watson and Crick made genetics not only necessary but also unavoidable in the understanding of Darwin's evolution by natural selection. In 1962, Watson, Crick, and Wilkins received the Nobel Prize for this landmark discovery of the DNA structure.

The development of the Watson–Crick structure of DNA led to the birth of molecular biology followed by the enunciation of the central dogma in biology. Molecular biology attempted to provide the molecular basis for everything in biology and biochemistry leading to the unity of life. Molecular biology perpetuated the reductionistic view of living systems: Reductionists attempt to understand a system by understanding its molecular components. Molecular biology also led to the development of a better understanding of diseases and their control by pharmaceuticals. The field of molecular biology ushered in by the Watson–Crick DNA structure led to the development of scores of Nobel Prize-winning concepts in biology, biochemistry, and medicine as discussed later in this book.

The coming of genomics marked the fourth major breakthrough in biology. Advances in genome sequencing and availability of human and several other genome sequences by 2001 provided the basis for the understanding of the uniqueness of humans in possessing certain distinctive DNA segments. Genomics also provides the basis for the understanding of variations among individuals as differences in DNA sequences. Furthermore, it provides molecular insight into the genetic basis for differences in our response to the same drug. The variation in individual DNA sequences is expected to provide the molecular understanding of our several complex traits, including behavior. DNA sequences also provide a better insight into the record of the evolutionary processes in an organism. Genomics is expected to provide a better understanding of a complex organism like humans after the elucidation of the roles of noncoding sequences (introns) of DNA. Understanding the roles of introns is currently a formidable task: It is believed that the elucidation of the roles of introns will add a new dimension to the understanding of biology.

The fifth breakthrough underway is the development of proteomics. This is bringing a better understanding of biochemical pathways and the roles of protein interactions. Above all, proteomics provides a clue to answering the big question of how a small number of genes can control several phenotypes in a complex organism like humans. A major conceptual scheme emerging from proteomics is that it is the number of interactions of proteins and not the number of proteins per se that is responsible for the myriad phenotypes in an organism.

The sixth breakthrough that is in making involves the science of synthetic genetics which would allow creation of new organisms by creation of entirely new genomes or by the manipulation of existing ones with the help of the techniques of molecular genetics, genomics, proteomics and bioinformatics.

Advances in genomics and proteomics in conjunction with bioinformatics have made it possible to realize the dreams of the chemists of the 20th century. These chemists wanted to decipher the amino acid sequences of all proteins to understand their functions. Proteomics has made it possible to determine the amino acid sequence of any protein. In addition, future advances in genomics and proteomics are expected to bring several revolutions in medicine and will make personalized medicine a reality. Advances in proteomics are expected to integrate the reductionistic views of Watson and Crick into systems biology to show how molecular parts evolved and how they fit together to work as an organism. The latter is expected to provide the ultimate understanding of biology.

1.1 INTRODUCTION TO PROTEOMICS

The term "proteome" originates from the words protein and genome. It represents the entire collection of proteins encoded by the genome in an organism. Proteomics, therefore, is defined as the total protein content of a cell or that of an organism. Proteomics is the understanding of the structure, function, and interactions of the entire protein content of an organism. Proteins control the phenotype of a cell by determining its structure and, above all, by carrying out all functions in a cell. Defective proteins are the major causes of diseases and thus serve as useful indicators for the diagnosis of a particular disease. In addition, proteins are the primary targets of most drugs and thus are the main basis for the development of new drugs. Therefore, the study of proteomics is important for understanding their role in the cause and control of diseases and in the development of humans as well as that of other organisms.

Proteins are encoded by DNA in most organisms and by RNA in some viruses. In all cases except RNA viruses, DNA is transcribed into RNA, which is then translated into a protein. In case of RNA virus, however, RNA is translated directly into proteins. Initially, it was thought that one gene makes one enzyme, which controls a phenotype. However, this view has undergone tremendous changes in the last several decades mainly because of the discovery of the split nature of eukaryotic genes, which involves RNA splicing, the occurrence of RNA editing, and the phenomenon of RNA silencing. The split nature of gene, RNA splicing, RNA editing, and RNA silencing are discussed later in this chapter.

In eukaryotes, the coding sequences of a gene called exons are interrupted by the noncoding stretches of nucleotides called introns. The exons are spliced after removal of introns within a gene continuously (referred to as cis splicing) or discontinuously (referred to as alternate splicing) or between exons of different genes leading to transsplicing. The different modes of splicing of exons and posttranslational modifications of proteins are responsible for the abundance of proteins in eukaryotic organisms. In humans there are approximately 23,000 genes and more than 500,000 proteins.

The findings of suppressor genes and the split nature of genes may present apparent contradictions to the one-gene–one-enzyme hypothesis. However, with the coming of central dogma (Crick, 1958, 1970, Watson 1965, Mattick 2003, Lewin 2004) in biology and elucidation of the genetic code (Leder and Nirenberg 1964, Khorana 1968), it is understandable how suppressor genes work. Thus, the mechanism of action of suppressor genes does not contradict the original ideas implicit in Beadle and Tatum's one-gene–one-enzyme concept to any extent as it appears superficially. In light of central dogma, it is understandable that certain genes or DNA segments may code for different proteins or that the coding section of protein in DNA is distributed across a huge expanse of DNA interrupted by the noncoding sequences. It has become obvious that the one-gene–one-enzyme concept applies only to genes that encode one polypeptide and not to genes that have a split nature and can code more than one protein. Thus, the one-gene–one-enzyme concept is limited to the nature of the gene itself, just as Mendelian rules of inheritance apply only to the genes located in the nucleus and not to the genes that are located elsewhere in the cell beyond the nucleus. Thus, the Mendelian inheritance pertains to the location of the genes, whereas the one-gene–one-enzyme concept is limited to the nature of the gene itself.

Obviously, what Beadle and Tatum suggested is not an axiom but a rule, and certain situations just represent exceptions to their profound rule. It seems that nature too has the British view of rule that "exceptions prove the rule." The history of science is full of such exceptions. The most glaring example of such an exception involves the central dogma in molecular biology described by Francis Crick, the codiscoverer of the DNA structure. Crick (1958, 1970) surmised that sequential information in DNA is transferred to RNA and then to protein from RNA and that the direction of this information transfer is fixed. However, later it was shown that RNA is reverse transcribed into DNA, and at times, messenger RNA (mRNA) is edited by the addition or removal of cytidine or uridine before its translation in to protein, which suggests that information in a DNA segment is not translated directly into protein as implicit in central dogma. This idea suggests that DNA makes RNA, which makes protein. Howard Temin and David Baltimore received the Nobel Prize in 1975 for demonstrating this reverse transfer of information from RNA to DNA. The other glaring example of such an exception includes the enzymes. It was James Sumner of the Cornell University who established that enzymes are proteins. Soon, enzymes became synonymous with proteins until Sydney Altman of Yale University and Thomas Cech of the University of Colorado showed independently that certain enzymes are made of RNA and not proteins. Sumner in 1946 and Altman and Cech in 1989 were awarded Nobel Prizes for their contributions to the science of chemistry. Thus, it seems that biology, like any other branch of science, is replete with instances of exceptions to the rules.

The Swedish scientist Berzelius (1838)[1] named certain naturally occurring polymers as proteins. The fact that enzymes are proteins was established by Sumner (1946). Later, Sanger (1958)established that proteins are made up of a sequence of amino acids. The fact that an enzyme and a substrate (or an antibody and antigen) require precise complementary fit in their structures, just like a hand in a glove, to interact with each other was established by Linus Pauling in the 1940s. In addition to Sumner (1946), both Pauling (1954) and Sanger (1958) received Nobel Prizes for their work in chemistry. Most proteins have enzymatic functions, but several of them such as actin and fibrinoactin are structural components of cells. Proteins are major constituents of muscle, cartilage, and bones. Proteins are also responsible for the mobility of muscle cells. Certain proteins serve as receptors for different molecules or work as immunoglobulins or antigens, or proteins can serve as allergens or participate in transport of various molecules, such as oxygen or sex hormones. Many proteins are hormones, such as insulin or human growth hormone (HGH), which control important

[1]The word protein was coined from the Greek word proteios first by Jöns Jakob Berzelius in 1838 in a letter to his friend.

metabolic functions in humans and other organisms. The three-dimensional structure and chemical modifications of proteins are important for the understanding of their functions in different capacities.

Gorrod (1909) first described certain human disorders as inborn errors of metabolism and implied the genetic basis of these diseases. However, it was the genius of Beadle and Tatum (1941) that led to the establishment of the fact that a protein is encoded by a gene. Working with, Neurospora, they showed that the synthesis of a substance in a metabolic pathway was impaired in a mutant. They showed that by disabling the gene controlling the enzyme that catalyzed a biochemical reaction in a metabolic pathway, the mutant developed nutritional requirements for that substance. Such mutants could not be grown on a minimal medium, but their growth was possible only when a particular substance was added to the minimal medium. For example, a mutant with impaired synthesis of arginine could not be grown on a minimal medium, but its growth was possible only when arginine was added to the minimal medium. This method was also used to map the biochemical pathways.

Beadle and Tatum (1941) called this conceptual scheme the one-gene–one-enzyme hypothesis. This hypothesis has been modified in various ways. However, despite several exceptions to this rule of one gene encoding one enzyme, the main tenets of the one-gene–one-enzyme hypothesis have remained the cornerstone of biology. This concept has been instrumental for the merger of chemistry with genetics and for the development of molecular biology. This theory provides the standard method to assign a function to a protein by creating a mutant and then showing which protein has a defective function or which function has been impaired in a particular protein. Because of this hypothesis, it was possible to analyze and study viral, microbial, plant, and animal genetics. This has been the basis for creating knockout mutations and for in vitro mutagenesis. This hypothesis has proven crucial for the analysis of any basic genetic mechanism, such as DNA replication, repair, and recombination, and for establishing the role of a protein in any metabolic pathway. Finally, this theory by Beadle and Tatum has led to advances in agriculture, animal husbandry, pharmaceutical sciences, and medicine. The one-gene–one-enzyme hypothesis has been the basis for the understanding and alleviation of human diseases and for the development of gene therapy.

The one-gene–one-enzyme hypothesis implied that a mutant must have altered the protein. Beadle and Tatum could not demonstrate the defective nature of the protein in their mutants because of the lack of technology at that time. However, this was demonstrated first at the biochemical level by Mitchell and Lein (1948, Mitchell, et al. 1948) and by Yanofsky (1952, 2005a,b) in tryptophan, which required mutants of Neurospora that lacked the enzyme tryptophan synthetase responsible for the synthesis of tryptophan. This concept was also demonstrated later at the molecular level by Ingram (1957) in the case of hemoglobin in persons who suffer from sickle cell anemia. Ingram showed that the sixth amino acid "glutamic acid," which is found in the hemoglobin of a normal person, is replaced by valine in the hemoglobin of a sickle cell person. This one change from glutamic acid to valine is the basis for the blood disorders in a sickle cell person. Later, many other mutants were shown to lack a protein altogether or possess proteins with altered amino acid(s).

The one-gene–one-enzyme theory also implied the correspondence in the ordered position of nucleotides in a gene with the position of amino acid in the protein encoded by that gene. This colinearity in the structure of a gene and that of a protein was demonstrated independently by Yanofsky et al. (1964) and by Sarabhai, et al. (1964), as discussed later in this chapter.

1.2 PROTEOME AND PROTEOMICS

1.2.1 Proteins as the Cell's Way of Accomplishing Specific Functions

The proteome is defined as the total proteins encoded by the genome of an organism. Proteomics is the science of describing the identification and features of the proteome of an organism.

The term "proteome" was first used by Marc Wilkins in 1994 (Wilkins 1996). An effort to describe the total proteins of an organism was made independently by O'Farrell (1975) and by Klose (1975). They developed what is called two-dimensional (2D) gel electrophoresis by running gel electrophoresis of proteins in two planes at right angles to each other (O'Farrell 1975, Klose 1975). This method separated a complex mixture of more than 1100 proteins of *Escherichia coli* into distinct bands of individual components on the gel. Later, the science of proteomics was revolutionized by the application of mass spectrometry in conjunction with genomics for the separation and identification of proteins on a large scale.

The genome of an organism is static in the sense that it remains the same in all cell types all the time. In contrast, the proteome of an organism is dynamic, because it differs from one cell type to another and keeps changing even in the same cell type at the different stages of activity or different states of development. A change in the proteome is a reflection of differential activity of the genes dependent on the cell type to express the protein needed for a particular function. For example, blood cells predominantly express the hemoglobin gene to produce the hemoglobin protein required for the transport of oxygen, whereas pancreatic cells largely express the insulin gene, which produces the insulin peptide required for the entry of glucose molecules into cells.

Thus, the differential expression of genes is required for the production of different proteins because each protein controls a distinct function. The function of many proteins is listed in Table 1.1. In addition, the protein profile of a cell can vary depending on the different kinds of modification of the same protein; such modifications of protein may involve acetylation, phosphorylation, glycosylation, or association with lipid or carbohydrate molecules. These modifications in proteins occur as posttranslational events and alter the function of proteins. One example is the mitosis activator protein (MAP) kinase protein controlling the mitosis; this protein is activated by phosphorylation to give MAP Kinase (MAPK), MAP kinase kinase (MAPKK), and MAP kinase kinase kinase (MAPKKK). The role of protein modification in the control of cellular activity is discussed later in this book.

1.2.2 Pregenomic Proteomics

The role of proteins as enzymes in controlling a cellular activity was known much before its structure was elucidated. The conceptual breakthrough in deciphering the structure of a protein as a linear array of amino acids came from the enunciation of the one-gene enzyme concept. This conceptual breakthrough was materialized by certain technical advances. The technical advances included the development of machines for the analysis of the amino acid composition and for the determination of the sequence of the amino acids in a protein. With the help of these machines, the structure of proteins was elucidated one protein at a time for several years. Later,

Table 1.1. Function of different proteins.

Function	Protein
1. Catalyst	Enzymes (more than 90% of proteins) Catalyze biochemical reactions in the cell
2. Transport	Hemoglobin (carrier of oxygen) Albumin (carrier of hormones)
3. Structure	Cartilage/bone proteins
4. Cellular skeleton	Actin, fibrinoactin
5. Hormone	Insulin, growth hormone
6. Antibody	Immunoglobulins
7. Antigens and allergens	Bacterial and viral proteins
8. Mobility/muscle movement	Myosin
9. Receptors	Receptor for cholesterol
10. Cell communication/signaling	Transduction proteins, junction proteins

the introduction of the methodology of the 2D gel and that of mass spectrometry facilitated the simultaneous resolution of the structure of several proteins at the same time. Understanding the structure of several proteins at the same time aided by mass spectrometry was moved forward with the coming of genomics and bioinformatics. The methods of genomics deciphered the nucleotide sequence of DNA/genes in the chromosomes of various organisms. The methods of bioinformatics involved the use of computers and several software programs for analyzing the bulk of the nucleotide sequence of DNA of an organism. Bioinformatics is also used for deciphering the amino acid sequence of a protein from the sequence of nucleotides in a DNA molecule.

1.3 GENETICS OF PROTEINS

A genetic approach to understanding protein structure and function was dictated by the one-gene–one-enzyme hypothesis. This concept implied that the structure and function of proteins could be understood by the comparison of the protein obtained from the wild type and from mutant organisms. In reality, it became a routine method to understand the role of a protein in any metabolic or developmental pathway. Following this dictum, the hemoglobin molecules from normal humans and from sickle cell patients were compared. The hemoglobin of normal individuals was found to be different from the sickle cell patients in the sixth amino acid. Normal individuals possessed glutamic acid at this position, whereas the sickle cell patient possessed valine (Ingram 1956, 1957). Thus, one change in amino acid completely altered the structure and metabolic role of hemoglobin (Figure 1.1).

1.3.1 One-Gene – One-Enzyme Theory

This theory proposed by Beadle and Tatum (1941) implied that the structure of an enzyme or a protein is controlled by one gene, in the sense that one gene encodes one protein. This theory became useful in understanding

	1	2	3	4	5	6	7	8
Hemoglobin A	Val–	His–	Leu–	Thr–	Pro–	*Glu*–	*Glu*–	*Lys*–
Hemoglobin S	Val–	His–	Leu–	Thr–	Pro–	*Val*–	Glu–	Lys–

Figure 1.1: A comparison of the N-terminal amino acid sequence in the beta chain of hemoglobin of normal and sickle cell patients.

the biochemistry of any metabolic pathway and the role of proteins that catalyzed the biochemical reaction at each step in that metabolic pathway. First, it became obvious that if an organism cannot grow without a supplement, such as a specific amino acid, nucleotide, or vitamin, then that organism is defective for the protein that catalyzes the biochemical reaction leading to the synthesis of that substance, which has become a nutritional requirement for its growth.

This led to the development of a methodology to identify mutants with a specific nutritional requirement and then the order of biochemical reactions in a metabolic pathway. Such an analysis of nutritional mutants revealed the presence of a different class of mutants. Among them, a class of mutants was found to require the amino acid ornithine or citrulline, or arginine for growth. Another group of mutants required either citrulline or arginine for growth, whereas the third group of mutants could grow only in the presence of arginine. The nutritional requirement of this last group of mutants was not met by adding ornithine or citrulline as a supplement to the growth medium when added alone or together. The nutritional requirements of these three groups of mutants suggested a metabolic pathway for the synthesis of arginine by the organism. Thus, this metabolic pathway involved the sequential steps of biochemical reactions involving the synthesis of ornithine from a precursor molecule and then the synthesis of citrulline from ornithine, and finally arginine from citrulline. Therefore, the metabolic pathway was established as follows: Precursor → Ornithine → Citrulline → Arginine. From this sequence of biochemical reactions in this pathway, it becomes obvious that the first group of mutants is defective in the step involving the conversion of the precursor into ornithine. Therefore, this group of mutants could use either ornithine, citrulline, or arginine for growth. The second group of mutants is defective in the step involving the conversion of ornithine into citrulline; therefore, its growth requirement could be satisfied by the addition of citrulline or argine but not ornithine. The third group of mutants is defective in the last step of biochemical reaction involving the conversion of citrulline into arginine, and thus, an organism could grow only when arginine is added as the supplement. Thus, the one-gene–one-enzyme concept became a useful tool in establishing the sequence of biochemical reactions in a particular pathway. This theory also implied that if the enzyme catalyzing the conversion of substance A into substance B is defective, then the molecules of substance A will accumulate in the organism. At times, the accumulation of this substance may cause a hazard to the health of mutant individuals. This is shown by the accumulation of phenylalanine in phenylketoneurics or the accumulation of homogentisic acid in infants who suffer from alcaptonuria. Such metabolic blockages occur in the metabolic pathway of phenylalanine–tyrosine pathways as a result of the specific enzyme defects, as observed in Figure 1.2. Such genetic defects were described as "inborn errors of metabolism" by Gorrod (1909). An accumulation of phenylalanine causes damage to the development of the brain in early stages of development, and it could lead to mental retardation. Now it is mandatory in the United States and other developed countries to screen babies after birth to check for phenylketoneuria by evaluating for an increased amount of phenylalanine in the blood. Phenylketoneuric babies are put on a special diet deficient in protein to manage the level of phenylalanine. After brain development is complete, these individuals are returned to a normal diet. However, a phenylketoneuric female must restrict the phenylalanine intake during pregnancy to allow the proper growth development of the infant's brain.

Later, this theory became useful in establishing the identification of a particular protein and its role in a biochemical step in the metabolic pathway by

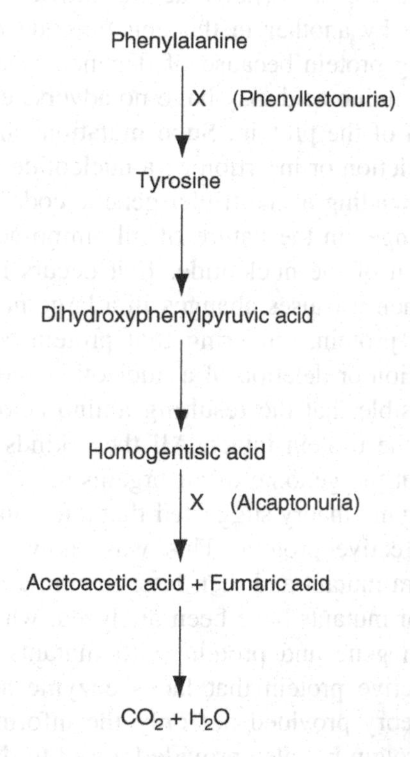

Figure 1.2: Consequences of a metabolic block in pheylalanine–tyrosine Defective phenylalanine hydroxylase can lead to the accumulation of phenylalanine, which can cause damage to brain cells and mental retardation in phenylketonuric babies. Another metabolic blockage caused by a defective enzyme can lead to alcaptonuria.

comparing the biophysical properties of the wild-type and mutant enzyme involved in the particular pathway. It was soon found that a mutant did not produce a particular protein, or produced a partial protein, or a defective protein with a different amino acid in a certain position in the protein. The occurrence of distinct classes of mutant proteins is consistent with the nature of changes that accompany a change in the genetic code. Such a change may involve the substitution of one nucleotide by another in the genetic code or a deletion or insertion of a nucleotide in the DNA sequence of the gene. A substitution of nucleotide in the genetic code may cause a nonsense, missense, or silent mutation in the protein. A nonsense mutation results from a change in the existing amino acid codon into a stop codon. A nonsense mutation that occurs in the beginning of a gene encoding the protein will make a small peptide or no protein at all. A nonsense mutation anywhere in the gene will yield a truncated protein of different lengths. A missense mutation that causes the substitution of one amino acid for another amino acid may alter the biochemical properties of the protein so that it is rendered inactive or partially active. However, such a substitution of one nucleotide by another in the genetic code may not cause any change in the resulting protein because of degeneracy of the genetic code or because a replaced amino acid may have no adverse effect on the overall structure and function of the protein. Such mutations are called neutral or silent mutations. A deletion or insertion of a nucleotide in the genetic code leads to a shift in the reading of the triplet genetic code. Such a frame shift mutation leads to changes in the nature of all amino acids from the point of insertion or deletion of the nucleotide. If it occurs in the beginning or middle of the gene, then it causes changes in a large number of the amino acids in the resulting protein, rendering that protein completely inactive. However, if the insertion or deletion of a nucleotide occurs toward the end of the gene, it is possible that the resulting amino acid changes may still leave the activity of the protein intact. All these kinds of mutations have been found to occur in the genome of an organism.

One-gene–one-enzyme theory suggested that a mutant would lack a protein or possess a defective protein. This was shown first in tryptophan requiring a Neurospora mutant and then later in similar mutants of *E. coli*. Currently, hundreds of mutants have been analyzed, which shows this one-to-one relationship in gene and protein with mutants always possessing no protein or a defective protein that lacks enzyme activity. Thus, one-gene–one-enzyme theory provided not only the informational role of the gene in encoding a protein but also provided a tool to dissect the biochemistry of any simple to complex processes in the living system by producing mutants and then comparing the biochemical changes in the mutant. No system has escaped the scope of this powerful tool.

1.3.1.1 Colinearity of Gene and Protein. The one-gene–one-enzyme concept of Beadle and Tatum (1941) provided the basis for colinearity in the DNA/gene and protein structures with a suggestion that the gene represents a sequence of nucleotides and the protein represents a sequence of amino acids. Avery et al. (1944) and Hershey and Chase (1952), by their transfection experiments in bacteria and bacterial viruses, established that genes are made up of DNA molecules. The fact that the gene is a sequence of nucleotides was shown by the correspondence between the genetic map of certain mutants with blocks of nucleotides. This colinearity between the DNA sequence of genes and the amino acid sequence of proteins was established by the study of missense mutants of *E. coli* (Yanofsky et al. 1964) or of nonsense mutants of a bacterial virus (Sarabhai et al. 1964). In both cases, the position of change in the genetic code corresponded with the position of amino acid change in the protein. Yanofsky et al. showed that a change in the early nucleotide sequence of a bacterial gene for protein A of tryptophan synthetase caused a corresponding change in the early amino acids in the protein. A change in the middle of the gene corresponded with a change in amino acid position in the middle of the protein. Similarly, a change in the end of a gene corresponded with a change in position toward the end of protein A of tryptophan synthetase. Sarabhai et al. (1964) showed that a virus produced truncated viral proteins; the size of the peptides corresponded with the length of the gene where the nonsense mutation occurred (Figure 1.3).

1.3.1.2 Protein as a Sequence of Amino Acids. The fact that a protein is a sequence an amino acid was directly established by the elucidation of the structure of insulin polypeptide as a linear sequence of different amino acids by Sanger (1958). Thus, insulin was the polypeptide or a small protein that was sequenced first by Sanger (1958). Ribonuclease A was the first full-size protein and an enzyme that was sequenced by

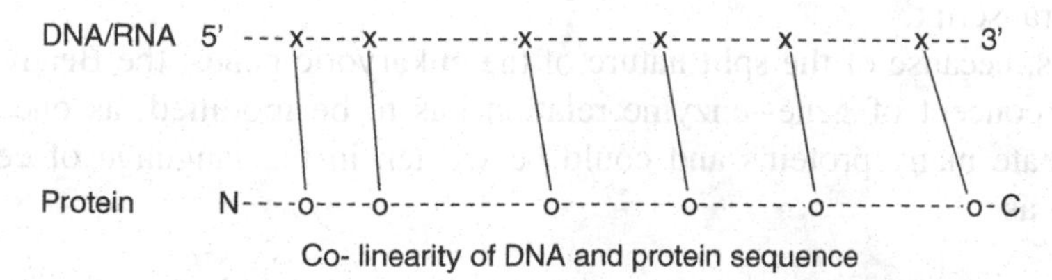

Figure 1.3: Colinearity of the DNA and protein sequence. The X represents the site of mutation in the gene/DNA as mapped by recombinational analyses. The O represents the position of altered amino acids in the protein coded by the gene. Vertical lines connect the position of changes in the gene and protein to show their one-to-one correspondence.

Stein and Moore (1972). However, the direct demonstration that a gene is a sequence of nucleotides was accomplished much later when the method for cloning of a gene and its sequence analysis became available. Proteins usually have four kinds of structure before a three-dimensional structure is assumed. These different structures are called a primary, secondary, tertiary, and quaternary structure (Figure 1.4). The linear sequence of amino acids in the proteins represents the primary structure. The secondary and tertiary structures originate from the folding of polypeptide on itself as a result of the interaction of the side groups attached to the amino acids. The quaternary structure results from the interaction of two or more fully folded polypeptides that interact with each other to give the protein structure.

The one-gene–one-enzyme concept did imply that the primary structure of the peptide determines the secondary, tertiary, and quaternary structure, and this was established by Anfinsen (1973) by an analysis of the mutant ribonuclease and by the study of chemical modification as well as the denaturation and renaturation kinetics of this enzyme (Anfinsen 1973).

1.3.1.3 One Gene—Many Proteins: Challenge to Proteomics.

The central dogma of biology suggests the direction of the flow of genetic information from DNA to RNA to protein is DNA → RNA → Protein.

In this scheme, the one-gene–one-enzyme concept of Beadle and Tatum is written as follows: One DNA → One RNA (Transcript or mRNA) → One protein This scheme holds well for the prokaryotic organisms, because in prokaryotic genes, the protein-encoding information is continuous and the transcript is directly translatable and equivalent to mRNA. However, it was soon found that many genes in eukaryotes have a split gene structure in that the protein-encoding segments (exon) in a gene may be interrupted by noncoding segments (intron). In view of the split nature of many eukaryotic genes, the transcript must undergo a process to remove the noncoding intervening sequences (introns) to make all coding segments or exons continuous to yield mRNA, which is translatable. The splicing of exons may occur in different ways and can lead to different kinds of mRNA from the same transcript.

Thus, because of the split nature of the eukaryotic genes, the Beadle and Tatum concept of gene–enzyme relation has to be modified, as one gene can create many proteins and could be written in the language of central dogma as

One DNA → One transcript → many mRNAs → Many proteins

It is of interest to note that the central dogma changed when it was found that RNA could be reverse transcribed into DNA. The central dogma is

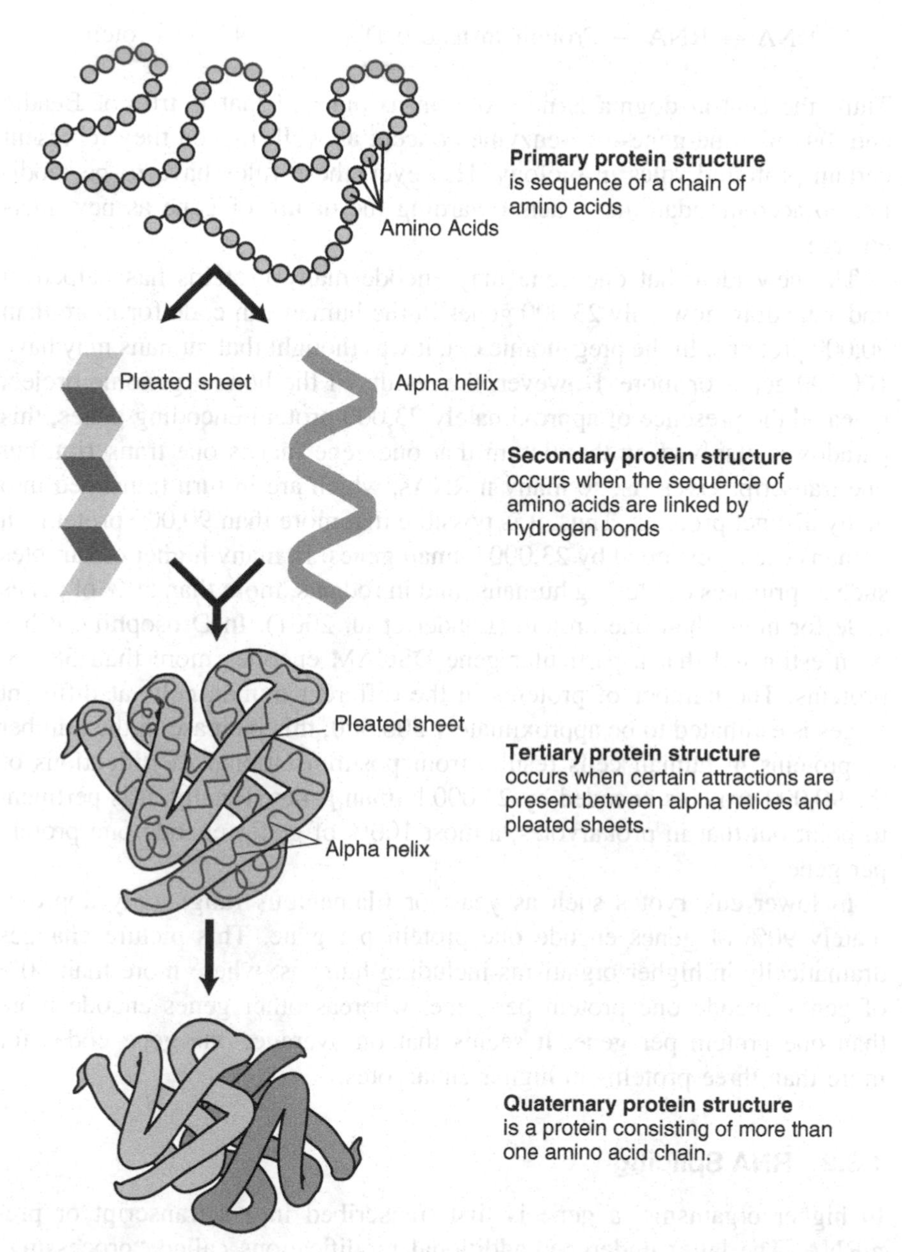
Primary protein structure
is sequence of a chain of amino acids
Amino Acids
Pleated sheet
Alpha helix
Secondary protein structure
occurs when the sequence of amino acids are linked by hydrogen bonds
Pleated sheet
Tertiary protein structure
occurs when certain attractions are present between alpha helices and pleated sheets.
Alpha helix
Quaternary protein structure
is a protein consisting of more than one amino acid chain.

Figure 1.4: Structure of protein with different levels of organization. Reproduced with permission of Darryl Leza of NIHGR/NIH.)

now depicted as

$$\text{DNA} \leftrightarrow \text{RNA} \rightarrow \text{Protein, instead of} \text{DNA} \rightarrow \text{RNA} \rightarrow \text{Protein}$$

Thus, the central dogma is no more an axiom and that is true of Beadle and Tatum's one-gene–one-enzyme concept as well. Indeed they represent certain profound rules in biology. However, these rules have to be modified to accommodate new facts regarding the nature of gene as new facts emerge.

The new idea that one gene may encode many proteins has helped in understanding how only 23,000 genes in the human can code for more than 90,000 proteins. In the pregenomic era, it was thought that humans may have 100,000 genes or more. However, the results of the human genome project revealed the presence of approximately 23,000 protein-encoding genes; this paradox is resolved by the dictum that one gene makes one transcript, but one transcript gives rise to many mRNAs, which are in turn translated into many distinct proteins. Thus, it is possible that more than 90,000 proteins in humans can be encoded by 23,000 human genes. In many higher eukaryotes such as primates (including humans) and in rodents, more than 50% of genes code for more than one protein (Lander et al. 2001). In Drosophila, it has been estimated that a particular gene DSCAM encodes more than 38,000 proteins. The number of proteins in the different human cells at different stages is estimated to be approximately 500,000; this increase in the number of proteins in human cells results from posttranslational modifications of the 90,000 proteins encoded by 23,000 human genes. Finally, it is pertinent to point out that in prokaryotes, almost 100% of genes encode one protein per gene.

In lower eukaryotes such as yeast or filamentous fungi, only approximately 90% of genes encode one protein per gene. This picture changes dramatically in higher organisms including humans, where more than 50% of genes encode one protein per gene, whereas other genes encode more than one protein per gene. It seems that on average, one gene codes for more than three proteins in higher eukaryotes.

1.3.2 RNA Splicing

In higher organisms, a gene is first transcribed into a transcript or pre-mRNA. The latter undergoes additional modifications called "processing" to produce translatable mRNA. The processing involves at least three steps. The first step includes a cap or the addition of novel guanosine nucleotide at the 5'end, and the second step includes a tail or the addition of a poly A nucleotides at the 3'end. The third step is the removal of intervening noncoding sequences called introns from the transcript. RNA splicing accomplishes the removal of introns and the joining of exons so that the different coding sequences in a transcript become continuous in the resulting mRNA. RNA splicing is carried out by a complex of RNAs and proteins organized into an organelle called a splicosome. A splicosome is as big as a ribosome and provides the platform on the surface of which the joining of exons and removal of introns are carried out. The two ends of an intron are recognized by certain concensus sequences such as GA at the 5'end and GU at the 3'end of the intron. During the process of RNA splicing, an intron loops out and is removed as a lariate structure with a guanine nucleotide as the tail bringing the neighboring exons together. Some introns are self-splicing and are removed without a splicosome. The RNA splicing of pre-mRNA occurs exclusively in eukaryotes. However, certain transfer RNAs (tRNAs) may undergo splicing in both prokaryotes and eukaryotes; their splicing is carried by out by certain enzymes without the involvement of splicosomes.

Eukaryotic pre-mRNA may be spliced out in different ways. First, the different exons of a particular pre-mRNA are brought together continuously by the removal of introns, which yields one translatable mRNA. For example, a pre-mRNA containing three exons and two introns will produce a mRNA after the removal of intons with all three exons together; such mRNA will produce a long protein on translation. Second, the different exons of this or similar pre-mRNAs may undergo alternate splicing, which yields several translatable mRNAs. For example, a pre-mRNA with three exons and two introns may undergo alternate splicing, which produces two different messages, one mRNA with exon one and exon two together, and other mRNA with exon one and exon three together. Thus, these two mRNAs will produce different proteins during translation. At times, certain exons of two different pre-mRNAs may be spliced together to yield different mRNAs. Such splicing that involves the exons of different pre-mRNAs is called transsplicing (Figure 1.5).

The process of alternate splicing is the major cause for the production of many proteins from one gene. The process of transsplicing causes the formation of one or more proteins from two genes. These two situations represent a major departure from the original one-gene–one-enzyme theory of Beadle and Tatum (1941). However, at the molecular level, it seems logical because enzymes or proteins are made up of modules encoded by the exons. Thus, nature has evolved ways such as alternate splicing and transsplicing to bring these modules together to produce a functional enzyme or protein.

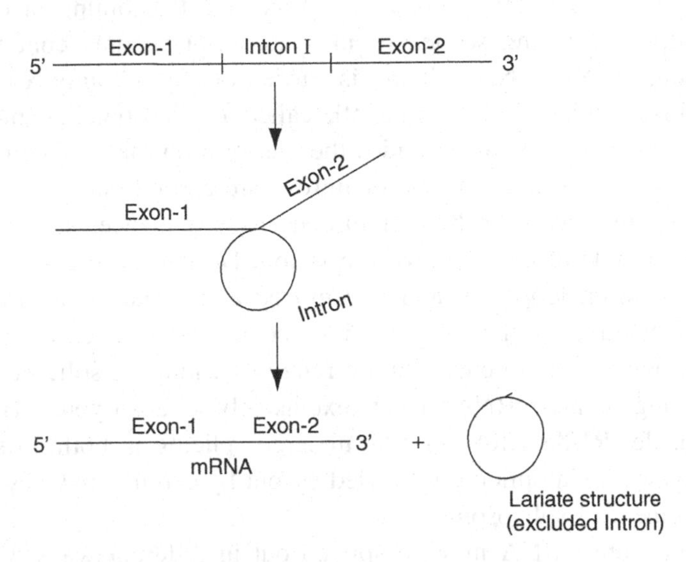

Figure 1.5: Removal of intron from a transcript.

1.3.3 RNA Editing

In addition to RNA splicing, the process of RNA editing is another factor that changes the nature of proteins. One gene may produce more than one functional protein through RNA editing. Thus, RNA editing can influence the proteomics of an organism. RNA editing involves the addition or deletion of cytidine or uridine nucleotide from the mRNA and causes a change in the nature of the codon in the mRNA before its translation. During RNA editing, the addition or deletion of a nucleotide is facilitated with the help of an RNA called guide RNA (gRNA). Often, organellar mRNA undergoes editing. In addition to insertion/deletion editing, RNA may undergo other kinds of modifications such as the conversion of cytidine into uridine or the conversion of adenosine into inosine by specific deaminases. These processes are called conversion editing. When adenosine is converted into inosine, it is translated by ribosome as a guanosine, thus, a CAG codon for glutamine becomes CGG after the conversion of adenosine into inosine, and it codes for arginine instead of glutamine. In addition to mRNA, tRNA, ribosomal (rRNA), and micro RNA (miRNA) may undergo editing. Usually, editing of tRNA leads to reading of a stop codon into leucine.

The process of RNA editing not only makes changes in the nature of protein but also presents an exception to the central dogma, it suggests because the direct transfer of information from DNA to RNA into protein.

RNA editing shows that at least in certain instances, proteins are made from information not present in the DNA sequence. Defective RNA editing has been associated with human cancer and with Lou Gehrig's disease, which is also called amyotrophic lateral sclerosis (ALS).

1.3.4 RNA Silencing and Proteomics

In recent years, an entirely new mechanism for gene control has been found to exist in plants, fungi, and animals. This approach involves the silencing of a particular gene-specific message by causing the degradation of mRNA. RNA silencing controls the expression of the resident gene (s), transgene(s), viral-induced gene(s), and transposons. It was discovered first in the petunia when a gene for anthocyanin was introduced to overexpress the color or pigment formation in the petunia flower. However, in such experiments, the expression of both the resident and the introduced transgene for color synthesis was suppressed, and the plant produced white flowers instead. This phenomenon was called as posttranscriptional gene suppression (PTGS). Later, similar gene suppression was found in the fungus Neurospora. It was determined that the introduction of a gene for orange color in Neurospora resulted in the transformants that were white or albino in color. This phenomenon for silencing a gene, such as the gene for pigment formation in Neurospora, is called quelling. It was found that the albino or white color Neurospora transformants did not produce mRNA specific for the color gene. It was also shown that only a part of the transgene containing only up to 130 nucleotides in length and not the entire gene for color was involved in the quelling of the resident gene. Such a transgene in Neurospora was found to quell or suppress the expression of resident genes even in another nucleus when a heterokaryon was constructed between the transformed and the wild-type strains of Neurospora. Later, Neurospora mutant strains were obtained that were defective in quelling; these mutants were called "quelling deficient" (qde). There are essentially three classes of such mutants in Neurospora. In Neurospora, qde-1 encodes for RNA-dependent RNA polymerase (RdRP), which is required for the synthesis of double-stranded RNA dsRNA such as miRNA or siRNA during gene silencing. The qde-2 gene encodes for the Piwi/Sting class of proteins related to a translation factor eIF2C. The Neurospora qde-3 gene encodes for a protein belonging to the group of WRN (Warner's syndrome) with RNase and DNA helicase functions similar to RecQ DNA helicase. The equivalents of Neurospora qde-1, qde-2, and qde-3 genes have been found to exit in different organisms, including Arabidopsis, worms (*Coenorobdytis elegans*), and fission yeast. Proteins belonging to the RdRP family have been well characterized from many plants, including tomato, wheat, petunia, and fission yeast, as well as from *C. elegans*. This protein is responsible for making a complementary copy of gene-specific mRNA. This copy of RNA hybridizes with mRNA to form a double-stranded RNA. The latter is degraded to smaller RNA fragments by an enzyme called Dicer, which is similar to RNase III ribonuclease. The RNA fragments then bind to an RNA-induced silencing complex (RISC) and cleave the mRNA specific to a particular gene, which causes gene silencing or suppression.

The role of dsRNA in silencing became obvious from the experiments with worms and Drosophila, and now it is found in mammalian cells as well. It was shown that the introduction of small pieces of dsRNA specific to a gene can cause the degradation of its mRNA, which leads to the suppression of the expression of that gene. Thus, RNA silencing can be used to manipulate the expression of genes in organisms and promise to serve as a great tool in the control of several human diseases, including cancer. Fire and Mello received the Nobel Prize in 2006 for elucidating the mechanism of RNA interference (Fire et al. 1998). It is known that certain infectitous agents, including viruses, trypanosomes, and intestinal parasites, cause havoc in humans because of their ability for antigenic variations. However, it is now known that certain intestinal parasites of humans such as *Giardia lamblia*, maintain their antigenic variation by the use of RNA interference (Prucca et al. 2008). Understanding this process may provide a clue to controlling several infectious diseases in human.

1.4 MOLECULAR BIOLOGY OF GENES AND PROTEINS

A gene is defined as a DNA segment or a stretch of nucleotide sequence that encodes a protein through the process of transcription and translation. However, a few genes make RNA that are not translated into proteins. These genes during transcription make rRNA and tRNA, which facilitate the translation of the transcripts from the protein-encoding genes. In prokaryotic genes, the coding segment of DNA is continuous and their transcripts are translated directly into protein without any modification. Thus, in prokaryotes, the transcript is synonymous to mRNA (i.e., the RNA that carries the information for making of a protein through the process of translation on ribosomes). The existence of mRNA in bacterial cells was demonstrated by Volkin and Astrachan (1957), and later the idea that mRNAs carry the information from DNA to ribosomes for translation into proteins was suggested by Brenner et al. (1961). Simultaneously, Marshall Nirenberg and H. G. Khorana elucidated the genetic codes and the mechanism of information storage and transfer as implied by the Watson-Crick structure of

DNA. Khorana and Nirenberg received the Nobel Prize in 1968 for their contributions. Even as early as the 1960s, the heterogeneous size of transcripts in eukaryotes was known. The eukaryotic transcripts were termed "heterogeneous nuclear RNA" (hnRNA) or premRNA. However, the myth about the heterogeneous nature of eukaryotic transcripts was elucidated by the discovery of the split nature of genes in eukaryotes. In the mid-1970s, it became obvious that some genes in eukaryotes have a split structure in which the coding segments of DNA called exons are interrupted by intervening noncoding DNA segments called introns. This conclusion was based on the results of heteroduplex mapping involving the hybridization of the DNA of a gene with mRNA and the visualization of the heteroduplex structure by electron microscopy. In such experiments, when the DNA of a gene was hybridized with the mRNA, certain DNA sequences appeared as loops (Sharp 2005). The appearance of these loops indicated the presence of the intervening noncoding sequences (introns) that were absent from the mRNA. Based on the results of these hybridization experiments, it was concluded that a transcript undergoes splicing events, which lead to an excision of the introns. Thus, the exons are made continuous and only then the message becomes translatable. These observations established a distinction between the structure of a transcript and its mRNA in the eukaryotes. Later, the presence of exons and introns in a gene was confirmed by comparing the DNA sequence of a gene and its mRNA for the chicken ovulbumin gene. With the completion of the genome projects of many organisms, the presence of exons and introns in a gene is readily established by identifying the occurrence of the conserved nucleotides at the exon–intron junctions.

Initially, it was thought that introns are simply excised out from a transcript by splicing, which makes the exons continuous in the mRNA. Later, it was shown that a particular transcript may yield many different mRNAs, which was facilitated by two different mechanisms called alternate and transsplicing. In alternate splicing, the exons are brought together in different combinations. For example, if there are three exons in a gene, an mRNA may contain exon 1 and exon 2, whereas another mRNA from the same gene contains exon 1 and exon 3 so that the two mRNAs will produce entirely different proteins with different amino acids in the C-terminal ends. These two proteins would have entirely different functions contolling different biochemical reactions in the physiology of an organism. Thus, depending on the number of exons, this method of alternate splicing may produce an array of mRNA for entirely different proteins.

It is suggested that in Drosophila, the DSCAM gene may produce more than 38,000 mRNAs encoding different proteins. Among an array of mRNA, not all mRNAs are translatable for a variety of reasons, including the presence of an early stop codon. Alternate splicing may be tissue specific and may produce proteins with a specific function. It is known that the Bcl-x gene makes a protein that controls programmed cell death or apoptosis. However, this gene makes two different mRNAs via an alternate splicing mechanism. A smaller version of mRNA produces a smaller protein Bcl-x(s), which promotes apoptosis and controls cancer, whereas a larger version of mRNA makes a larger protein that suppresses apoptosis and supports the growth of cancer.

Alternate splicing may involve exon skipping or intron retention. Exon skipping is commonly found in higher eukaryotes. During exon skipping, a particular exon is skipped during splicing. There are several examples of exon skipping, which is used to produce different versions of tropomyosin specific for skeletal muscle, smooth muscle, and brain cells. Exon skipping is found in Drosophila for the control of sex development. Drosophila has a sex lethal gene called sxl; when exon 2 is skipped during splicing, a female-specific sxl protein is produced, which binds with all subsequent transcript of the same gene, causes exision of exon 2 from all mRNAs, and leads to the development of female flies. However, if the male-specific exon 2 is retained in the first round of splicing, it leads to the production of the male-specific sxl protein and causes the development of male flies.

Intron retention results in the production of mRNAs and their encoded proteins of different lengths. Intron retention is commonly found in plants and lower multicellular organisms.

The mechanism of alternate splicing always involves one transcript. As opposed to alternate splicing, transsplicing involves the splicing of exons of two transcripts produced by the same or distinct genes (Figure 1.6). Transsplicing is commonly found in worms such as *C. elegans*. There is some indication that transsplicing may occur in human brain cells.

On average, a protein-encoding gene in humans is roughly 28,000 nucleotides in length, contains approximately 8 exons of 120 nucleotides or more, and contains approximately 7 introns varying in size from 100 to 100,000 nucleotides. Introns are usually several times of exons in length. A human gene on an average produces 3 mRNAs via alternate splicing.

Splicing is facilitated by splicosomes that consist of more than 100 proteins and five small nuclear (sn) RNAs (snRNAs). Certain regulatory proteins called "splicing regulator (SR) proteins" bind to a particular nucleotide sequence in the exon called the exon splicing enhancer (ESE) and recruit splicosomes. The exon may contain an exon splicing suppressor (ESS) sequence, which prevents the splicosome from splicing.

Defective splicing may cause diseases in humans. More than 15% of mutations that cause diseases in humans result in defective splicing. Defective splicing may result in mutations that alter the splice site or the components of splicesosomes, or it may change factors that control splicing.

1. Cis splicing or intramolecular splicing

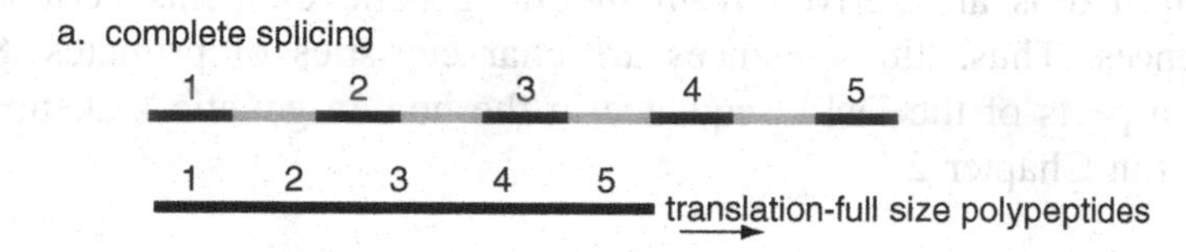

b. Alternate splicing

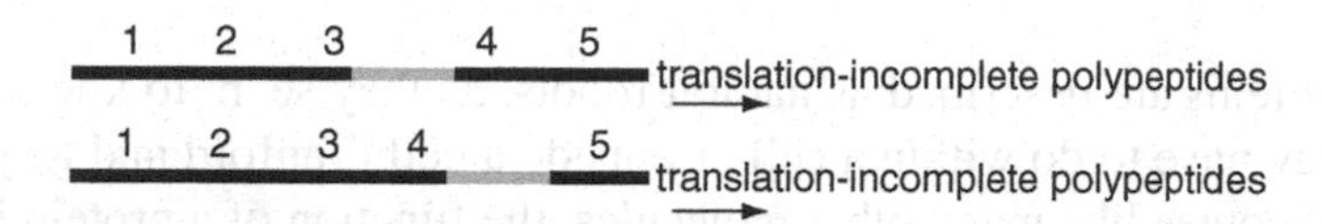

2. Tran splicing or intramolecular splicing

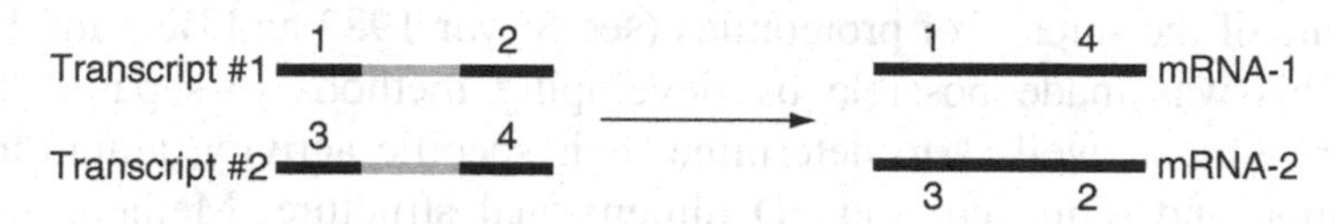

Figure 1.6: Different kinds of splicing of transcripts. (Reproduced from Mishra, 2002 with permission of John Wiley & Sons.)

Many human diseases including cancer may involve mutations that cause defective splicing (Faustino and Cooper 2003). Some genes in which a mutation is known to cause defective splicing and human diseases include BRCA1; BRCA2, HGH, cystic fibrosis; spinal muscular atrophy (SMA), myotonic dystrophy (MD), Wilms tumor suppressor associated with Frasier syndrome (WT1), and many more.

Alternate splicing is the major source of the abundance of proteins in higher organisms. Alternate splicing not only increases the number of proteins but also alters the nature of the protein by insertion and removal of codons in the resulting mRNA. It may also change the reading frame of the mRNA. It could cause termination of protein synthesis by introducing a termination codon in the mRNA. Alternate splicing may control gene expression by changes in the regulatory elements that affect mRNA stability and the translation process.

Alternate splicing has a great effect on speciation as is revealed from the understanding of the genome sequences of humans and mice.

Both have the same number of genes and even share the same exons and introns in many genes. However, it is believed that approximately 25% of the exons that undergo alternate splicing are specific to humans and are different from mice. Likewise, primates have primate-specific alternately spliced exons that set in the evolution of primates. It seems that the primate-specific exons are derived from mobile genetic elements containing alu sequences. Thus, alu sequences are characteristics of primates. Some of these aspects of the DNA sequence in the human genetic makeup are discussed in Chapter 2.

1.5 PROTEIN CHEMISTRY BEFORE PROTEOMICS

Proteins are described as natural robots, as they seem to know exactly what they have to do within a cell or outside a cell (Tanford and Reynolds 2004). Of course like many other molecules, the function of a protein is determined by its structure. As mentioned, proteins may function in many ways (see Table 1.1). Much of the basic biochemistry of protein was established before the coming of the science of proteomics (see Stryer 1982 and Bell and Bell 1988). This was made possible by developing methods to separate and purify proteins, as well as to determine their specific activity, amino acid composition and sequence, and 3D dimensional structure. Methods were also developed to characterize other physical and biochemical properties, including their regulation and artificial synthesis.

1.5.1 Separation and Purification of Proteins

Proteins were separated from each other during preparation of the cellular extract. Several methods are available to extract proteins from cells or tissue. Proteins are separated by the precipitation in different concentrations of ammonium salts usually in a stepwise manner. Partially purified proteins are separated based on differences in their molecular weights and charges. A small amount of proteins is usually purified based on differences in the molecular weights by ultracentrifugation in a sucrose gradient. Alternatively, they are separated by the method of gel filtration, which acts as a molecular sieve to separate protein molecules based on their sizes. Sepharose (Sephadex; Sigma-Aldrich, St. Louis, MO) is commonly used as a molecular sieve to separate protein molecules of different sizes. Proteins are also separated based on their net positive or negative charges by ion-exchange chromatography. Celluloses such as carboxymethyl (CM) cellulose and diethylaminoethyl (DEAE) cellulose are used in such an ion-exchanger matrix. Several other chromatography methods have been developed that separate protein molecules by their sizes as well as by their charges. Besides chromatography on a solid matrix such as sepharose methods for liquid and high-performance liquid chromatography (HPLC) also have been developed. A large number of proteins has been purified to homogeneity. Several proteins have been crystallized, and their three-dimensional (3D) structures have been determined.

In addition to different methods of chromatography, several methods of electrophoresis have been developed to separate protein molecules both based on their mass as well as on their charges on a regular gel or on a capillary gel by applying an electrical field. Sodium dodecyl sulfate (SDS) is added to the gel matrix to separate proteins of different molecular weights. SDS is highly negatively charged, and in its presence, all proteins in a mixture become equally negatively charged. Thus, during electrophoresis in the presence of SDS, all proteins move in an electrical field based on their molecular sizes and not on their charges. Smaller proteins move much faster than larger proteins during electrophoresis in a gel that contains SDS. A mixture of proteins also can be separated based on net charges by electrophoresis in a gel that contains a mixture of ampholine of different isoelectric points (pIs). These two methods of eletrophoreses in SDS gel and ampholine gel are combined so that a protein mixture is first run in SDS gel and then in ampholine gel to separate them based on molecular sizes and electrical charges. This method is called 2D gel because it separates proteins based on sizes and charges when run in two planes at right angles to each other. During electrophoresis, proteins are obtained as separate spots visualized by coloring with a dye. 2D gel was first used to separate more than 1100 proteins of *E. coli* simultaneously on one gel. The ability of 2D gel to separate the entire protein content of an organism and to provide information about them in one attempt ushered in an era that marked the beginning of the science of proteomics.

1.5.1.1 Specific Activity of Proteins. The specific activity of a protein is defined as the activity of a protein preparation per milligram of that protein. The activity of protein is usually determined as the enzymatic activity, as the ability to bind to a ligand, or as its biological activity. The specific activity of a protein increases with the increase in the purification of a protein. There are several methods to determine the amount of protein in a preparation. The simplest way is to measure the absorption at 280 nm of light. In addition, several colorimetric methods are available, of which the Lowry method and the Bradford method are commonly used (see Bell and Bell 1988).

1.5.1.2 Molecular Weight Determination. The molecular weight of a protein is an important criterion. It provides the idea about the relative size of the protein molecules. The molecular weight is traditionally determined by ultracentifugation or by chromatography through a matrix such as Sepharose or by the mobility of protein molecules in SDS gel on electrophoresis with reference to known protein markers.

1.5.1.3 Amino Acid Composition. Proteins comprise 20 different kinds of amino acids. It is important to know the relative abundance of these component amino acids in a protein. Knowledge of the numbers of different amino acids is also important in determining the sequence of the amino acids in a protein molecule. To determine the amino acid composition of a protein, it is hydrolyzed in 6 M HCl for a few hours and then separated by electrophoresis or by chromatography. The individual amino acid spots on an electrophoretogram are dyed with ninhydrin to facilitate its visualization. The number of amino acids in each spot is determined by colorimetric because the intensity of dye in each spot is related to the number of amino acids. Alternatively, the amino acids separated by chromatography as eluents are dyed with fluorescent dye, and the number of amino acid in a particular eluant is again determined spectroscopically as the number of amino acid is proportional to the amount of dye absorbed by the amino acids. The whole process is automated, and the commercially available machine called an amino acid analyzer determines the amino acid composition of a protein within a few hours. The amino acid analyzer was first developed at Rockefeller University in New York City.

1.5.2 Amino Acid Sequence

The sequence of an amino acid in a protein is determined sequentially from the N-terminus. The N-terminus amino acid is identified by the Edman degradation reaction developed at Rockefeller University and later automated in Melbourne, Australia, by Edman and his collaborators (1950, 1967).

To sequence a protein, it is usually fragmented into peptides of approximately 50 amino acids by cyanogens bromide cleavage or by tryptic digestion. Peptides are first separated from one another. A particular peptide is then adsorbed on to a solid surface such as glass fiber coated with cationic polymer polybrene. An Edman reagent phenylisothiocyanate (PTH) is added to the adsorbed peptide in a basic buffer solution of trimethylamine. In this solution, PTH reacts with an amino group of the N-terminal amino acid, which is then selectively separated from the peptide by the addition of an anhydrous acid. The modified N-terminal amino acid isomerizes into phenylthiohydantoin. This is washed off and then identified after chromatography. The cycle is then repeated to determine the next N-terminal amino acid in the remaining peptide that is adsorbed on to glass fiber coated with polybrene. The method of Edman degradation is elegant but riddled with certain limitations. The method will not work if the N-terminal amino acid is blocked or buried in the bulk of protein.

1.5.3 Chemical Synthesis of Protein

Chemical synthesis of protein has a long history. Synthesis of the first dipeptide glycylglycine was accomplished by Emil Fischer in 1901. Later, he synthesized octadecapeptide with a different amino acid sequence consisting of 15 glycine and 3 leucine amino acid residues. During such synthesis of peptide, he could not control the sequence of amino acids. An important advancement in this direction was made by Bergmann and Zervas (1932) in Germany by introducing the methods for protecting the amino group. In 1935, both Bergmann and Zervas joined Rockefeller University and trained several protein biochemists including William Stein and Standford Moore, who were awarded the Nobel Prize in 1972 as mentioned earlier in this chapter. Using this strategy of Bergmann and Zervas, an octapeptide hormone oxytoxin was synthesized in 1954 by du Vigneaud et al. Vincent du Vigneaud won the Nobel Prize in Chemistry in 1955 for the synthesis of oxytoxin. However, these methods for chemical synthesis in the solution phase were time consuming. A major stride in the chemical synthesis of protein was made by Merrifield in 1963 at Rockefeller University by developing solid-phase synthesis. In this method, an amino acid is attached to an insoluble support through its carboxyl end and then is reacted by another amino acid with an activated carboxyl group but is protected by an alpha amino group. The amino group of the dipeptide is then deprotected by the removal of the protecting group at the amino terminal and then reacted by a third amino acid with a protected amino group and a activated carboxyl group, which leads to the synthesis of the tripeptide. This process of protection, activation, and deprotection is continued in a cyclic manner until the synthesis of the entire peptide or protein is completed. During such chemical synthesis, it is important to protect certain reactive side chains of the amino acid. At the completion of the chemical synthesis, all protected groups are deprotected and then the peptide is cleaved off the solid support. Such peptides or proteins are then examined for the biochemical and biological properties to demonstrate their identity with the naturally synthesized protein. Merrifield used this method to synthesize the first enzyme ribonuclease A (RNaseA). The method for the chemical synthesis is now automated completely. The entire peptide synthesis is carried out by a machine developed at Rockefeller University.

It is important to note that the knowledge of the sequence of amino acids in RNaseA was crucial in the chemical synthesis of this enzyme by the Merrifield group. It is also important to note that the biosynthesis of a protein inside a cell always occurs from the N-terminal amino acid, whereas in the chemical synthesis, the peptide chain grows from the C-terminal amino acid, which is first attached to an insoluble solid support. It is important to note a long-chain protein like RNaseA is first synthesized in vitro as several component peptides and then they are ligated to yield a full-length protein. Usually, an acid-sensitive tert-butoxycarbonyl (Boc) group or a base-sensitive 9-fluorenylmethyloxycarbonyl (Fmoc) group is used to protect the alpha-amino group of the amino acid to be added to the growing chain of peptides during the chemical synthesis. A detailed method of the chemical synthesis of protein is discussed elsewhere (Nilsson et al. 2005).

1.5.4 Protein Engineering

Our ability to synthesize proteins in vivo and in vitro has led to the development of protein engineering technology. Using this technology, proteins of interest with certain desirable properties are produced in abundance. The process of protein engineering uses two different methods, which are not mutually exclusive. In reality, most laboratories use both methods for protein production. The first method is called "rational design." This requires a complete knowledge of the protein structure, which was difficult in the preproteomic era but has become readily available after the development of proteomics. This method uses site-directed mutagenesis and is a cost-effective method. The second method is called "directed evolution". This method mimics the process of natural evolution because proteins of different kinds are produced by random mutgenesis and then screened to select one with desired features. At times, DNA encoding different proteins are spliced to construct an end-product that combines the desirable features of different proteins. The major drawbacks of this strategy are twofold, one that it is a laborious method that involves several constructs and second that it requires high throughput, which is not possible for certain proteins.

1.5.5 Crystal Structure

The analysis of the structure of the protein crystal provides insight into the 3D structure of the protein with respect to the location of every atom one after another in the amino acid string of the protein. The 3D structure of the protein is usually revealed by the X-ray defraction pattern of the protein crystal. The X-ray defraction pattern is generated by scattering the X ray by the electrons in the atom when a beam of X ray is shined on the protein crystal. The X-ray pattern is then subjected to analysis by Fourier transformation to generate the 3D structure. An analysis of the 3D structure protein was conducted by John Kendrew and by Max Perutz in the Cavendish laboratory some time in the early 1940s. It took them almost 22 years to determine the 3D structure of a small protein myoglobin (Kendrew 1961) and that of hemoglobin (Perutz et al. 1960), for which both Kendrew and Perutz received the Nobel Prize in 1962. After their work, the 3D structure analysis of protein progressed slowly. By 1990, the structure of less than 100 proteins was revealed by determining the X-ray defraction pattern of the protein crystals.

The whole process was sped up by advancement of a new technique called mad, in which a synchatron was used to beam X ray on the protein crystals. This technique readily provided data about the phase of defraction. To generate the 3D structure, both information regarding the amplitude and phase are required.

An earlier phase was determined by the X-ray defraction of protein crystals containing heavy metals at different positions that required the comparison of several X-ray defraction patterns. The process of the 3D structure analysis was sped up by the advances in computing power as well.

In addition to X ray, nuclear magnetic resonance (NMR) is used to determine the 3D structure of small proteins in solution. NMR is good for proteins that cannot be crystallized. NMR also yields the 3D structure of proteins in dynamic state because protein molecules are in solution unlike the 3D structure of a protein crystal.

1.5.6 Active Site and Regulation of Proteins

One important aspect of proteomics is to understand the function of a protein. This is particularly crucial for understanding the role of protein in causing diseases and in developing drugs. Proteins act in several ways in a cell. Most proteins act by catalyzing a biochemical reaction or by binding with certain molecules including other protein molecules. A protein usually contains an active site as a part of its structure; with the development of proteomics, the active site of an enzyme can be determined by bioinformatics using computing software. The active site binds with a substrate during enzymatic reaction and then catalizes the reaction. Two models are used to explain the binding and catalysis of a substrate. The first model is called the "lock and key model," and the second model is called the "induced fit" model. In the first model, the active site and the substrate have a lock and key relationship, which accounts for their specifity. In the induced fit model, the active site is not a rigid structure and suggests certain flexibility in the active site induced by the binding of a substrate. Molecules that mimic the structure of the substrate can bind with the active site and can block binding with the substrate; this is the basis for enzyme inhibition by certain drugs and the regulation of enzymes, as discussed below.

Such sites may bind with other proteins or with certain other molecules in the case of nonenzyme proteins. Certain drugs that bear a similarity to the structure of the substrate may bind with the active site of protein and may inhibit the enzymatic activity of the protein by obstructing its interaction with the natural substrate. Such inhibitors are known as competitive inhibitors, because they compete with the substrate for the active site of the protein. Unlike the molecules that bind with the active site of a protein, certain other molecules bind with the protein at a site other than the substrate binding site and bring a conformational change to the structure of the protein such that it cannot bind with the substrate anymore. Such inhibitors are known as allosteric inhibitors and as noncompetitive inhibitors, because they do not compete with the substrate for binding with the active site. The competitive and noncompetitive inhibitors are easily distinguished by Michaelis-Menton kinetics. Such kinetic analysis is carried out by plotting 1/V against 1/S, where S represents the substrate concentration and V represents the velocity of the biochemical reaction (see Bell and Bell 1988), which generates a straight line. The allosteric regulation of protein was established by Jacob and Monod (1964) in Paris. With the creation of appropriate mutants of *E. coli*, Jacob and Monod established the role of an allosteric protein or a repressor protein involved in the control of transcription of the Lac operon in this bacterium. Both Jacob and Monod received the Nobel Prize for this work in 1965.

1.5.7 Signal sequence and Protein Targetting

Proteins are synthesized on ribosomes and then move from their place of syntheses to the different parts of the cell to take residence there and to function in different ways. In 1970s, Gunter Blobel of Rockefeller University identified about 15 amino acids long sequences in different proteins that targets these proteins to their destinations into the cell wall, cell membrane and different organelles including nucleus, nucleolus, golgi bodies, mitochondria, chloroplasts and periosomes or to the exterior of the cell. These Signal sequences are like postal codes that help deliver letters to final destinations. The signal sequences are usually located on the N-terminus of the protein. The signal sequences are usually cleaved by a protease after the transport of the proteins. Proteins transported to different locations usually carry signal sequences of different amino acid composition for example proteins that are destined to endoplasmic reticulum consist of signal sequences of 5-10 hydrophobic amino acids at the N-terminus whereas the signal sequences of those proteins being transported to nucleus contain plus-charged amino acids within the peptide. The mitochondrial targeting signals contain alternating sequence of hydrophobic and plus-charged amino acids. The proteins being targeted to peroxisomes usually carry a signal sequence of three amino acids on the C-terminus. These proteins destined for transport are usually unfolded and are escorted by a chaperon protein. After the transport is complete the unfolded proteins are folded to assume tertiary structures with the help of a chaperon protein. At times protein can find its destination upon glycosylation i.e. acquisition of carbohydrate moiety.

A genetic defect in this protein transport leads to a number of human diseases. Therefore the understanding of protein transport has been instrumental in understanding these human diseases and may provide clue for their therapy. Blobel was awarded Nobel Prize in 1999 for his work elucidating the mechanism of protein transport in the cell.

The views regarding the general distribution of proteins and enzymes inside the cell, have changed over the years. Earlier it was thought that enzymes are randomly distributed in the cytosol and the enzymatic reactions happened by the chance meeting of an enzyme with the substrate molecule. Contrary to this view, the enzymes of same or related metabolic pathways are localized together and not randomly distributed in the cytosol. They occur together in the vicinity of each other by virtue of certain structural similarities which help them in recognizing each other. This view is also supported by the study of protein-portein interactions discussed in Chapter 5.

1.5.8 Intein

Inteins are segments in a protein that are self-exised out, followed by the joining of remaining segments called the exteins. After the removal of intein, the N-terminal and C-terminal exteins are joined by a peptide linkage as soon as the peptide is synthesized from the mRNA. Inteins in proteins are like introns in genes; intein must be removed to provide a functional protein just as an intron must be removed from a transcript to give a translatable message. Currently, more than 200 inteins have been described from different proteins; a data bank of inteins is available. Inteins are usually 100–800 amino acids in length. Some inteins may be derived from two genes encoding them; for example, the dnae DNA polymerase (dnaE) of cyanobacteria contains two segments, an N-intein segment of 123 amino acids and a C-intein sement of 36 amino acids. The two segments are encoded by two separate genes dnaE-n and dnaE-c for the alpha subunit of the DNA polymerase III. This is equivalent to transsplicing in the case of the genes. The gene-encoding inteins usually carry an endonuclease that helps in the propogation of inteins. Inteins have been found in all forms of life, including archea, bacteria, and eukaryotes. Inteins have been used in different ways, such as protein engineering and marking a protein for NMR characterization. Inteins may provide a useful tool to develop a drug that can stop the removal of intein from a protein, which renders that protein nonfunctional and, therefore, responsible for the cause of a disease.

1.5.9 Unstructured Protein

Now it is established that there are two classes of proteins in the living systems: one class with an ordered structure and a second class without any ordered structure, intrinsically unstructured, or unordered (Dyson and Wright 2005).

The structure of the first group of proteins was well established before the development of proteomics. All proteins are known to assume several levels of organization, such as the primary, secondary, and tertiary structures. Several proteins have another level of organization called the quaternary structure. The sequence of amino acids in a protein represents the primary structure. The secondary structure represents the coiled structure of the protein because of the folding in the primary structure based on interactions of the amino acids, particularly their side chain among themselves. The tertiary structure of a protein is its 3D structure based on the complete folding of the polypeptide on itself. The tertiary structure determines the final shape of the protein and its activity. Many proteins, after assuming a tertiary structure, interact either with themselves or with another protein to assume a quaternary structure. Proteins with identical peptides as subunits in the quaternary structure are called homomers, whereas those with different peptides as a subunit in the quaternary structure are called heteromers. Hemoglobin that contains two alpha chains and two beta chains is an excellent example of a protein with a quaternary structure. No proteins exist in its primary structure as a straight stretch of peptide as soon as the proteins are synthesized because of the immediate biochemical interactions among the different amino acids in the stretch of polypeptide. The secondary structure is usually determined by circular diachroism or even by gel filtration. The 3D structure of a protein with its tertiary or quaternary structure is best determined by X-ray crystallography or by NMR spectrometry as discussed in Chapter 3. The different structures of a protein can be determined several ways. Not all proteins assume a 3D structure. It is estimated that more than 35% of proteins found in living systems have no intrinsic structure, as they lack a tertiary structure. The proteins are called intrinsically unstructured proteins (Dyson and Wright 2005). These proteins usually lack the bulky hydrophobic amino acids in their primary structure. These proteins exist as random coil chain and are short lived. They usually perform regulatory functions, such as controlling the regulation of cell cycle, regulating transcription and translation, and signaling pathways. The study of these proteins is in infancy, but it is expected to throw much light on how the function of proteins is regulated.

1.5.10 Protein Misfolding and Human Disease

In addition to the random coil structure of the intrinsically unstructured proteins, other proteins also may assume such an unspecified structure because of mutation or other changes in the proteins. Proteins in several neurodegenerative and other diseases are altered in their structure. They lack the usual secondary and tertiary structure of the protein. They usually result from the misfolding of proteins. As mentioned, Anfinsen established that the secondary and tertiary structures are controlled by the primary structure of the protein. He demonstrated that the enzyme ribonuclease unfolds under denaturing conditions, such as during the addition of urea to the solution that contains the protein. Unfolded ribonuclease loses its enzymatic activity. However, as urea is removed from the protein solution by dialysis, ribonuclease starts folding and assumes a completely folded structure with a secondary and tertiary structure and the full restoration of its enzymatic activity. Anfinsen won the Nobel Prize in 1972 for this work. However, many proteins cannot fold, as they are synthesized inside the cell and remain misfolded. The existence of such misfolded proteins causes several diseases in humans. Certain neurogenerative diseases, including Alzheimer disease, Creutzfeldt-Jakob disease, Kuru, and mad cow disease, result from the misfolding of proteins.

In the 1960s, some proteins that cause mad cow disease were characterized as infective protein molecules. Because they possessed proteins exclusively, these were called prions, which are analogous to virions; the nucleic acid contained the infecting particles. Later, Prusiner, who characterized the prions as protease-resistant proteins (PrPs), was awarded the 1997 Nobel Prize in Physiology and Medicine for his work. The gene for the normal PrP after mutation causes prions in which the mutant PrP cannot fold properly. Later, these misfolded PrPs were shown to be infective protein molecules that perpetuate by causing the misfolding of proteins that otherwise would have exited as normal properly folded proteins with full biological activity. Several other diseases such as cystic fibrosis also result from the misfolding of a protein called the cystic fibrosis transmembrane conductance regulator (CFTR). These misfolded proteins remain in the random coil position and lose the biological activity required for the transport of chloride ions. However, in Alzheimer disease, they become sticky and form the characteristic plaques of beta sheets in the brain of the patients.

When a protein molecule gets misfolded for some reason, it is usually degraded. However, sometimes it escapes degradation and then it acts like a cheparon and causes the misfolding of the other protein molecules. This is the basis of the so-called infectivity of prions that causes mad cow disease and Creutzfeldt-Jakob disease. Once the healthy cow is exposed to misfolded protein or prions, it leads to the misfolding of the other proteins in the brain cells, which causes the disease.

When prions were first discovered, they were considered as infectious protein particles, and for a while, it was thought that the prions acted as a proteinaceous infecting agent parallel to the infective viral DNAs/RNAs. This view presented a challenge to the age-old dogma that only nucleic acids acted as genetic material. However, with the understanding that the formation of new prion particles is induced by the misfolding of other naturally occurring proteins, the myth of a protein as genetic material has been resolved. The prions cannot replicate and thus do not code for the daughter prions. Instead, these prions recruit new prion particles by inducing misfolding of the newly synthesized proteins encoded by the host genome. For example, in cystic fibrosis, a misfolded CFTR protein leads to the misfolding of other CFTR proteins, and the cell loses its normal function. It is shown that the normal PrP controls the long-term memory in mammals. Prions or prion-like particles have been found to exit in yeast and fungi-like Podospora, where they control different phenotypes in the organisms that harbor them. Thus, the understanding of proper folding of protein is crucial in knowing the cause of these diseases and their treatments. Now, a yeast heat shock protein Hsp40/YdjI has been identified that suppresses the aggregation of misfolded proteins and helps in refolding misfolded proteins. It seems to recognize certain repeat sequences as a consensus motif in the protein. The *E. coli* protein Dnaj is homologous to yeast protein. This protein may be useful in understanding human diseases that involve the misfolding of proteins.

REFERENCES

Anfinsen C. B. 1973 Principles that govern the folding of protein chains. Science 181, 223.

Avery, O. T., C. M. MacLeod, and M. McCarthy. 1944. Studies on the chemical nature of the substance inducing transformation of pneumococcal types I induction of transformation by DNA from Pneumococcus type III. J. Exp. Med. 79, 137

Ayala, F. J. and J. A. Kiger. 1984. Modern Genetics, Second Edition. Menlo Park, CA: Benjamin Cummings.

Beadle, G. W. and E. L. Tatum. 1941. Genetic Control of Biochemical Reactions in Neurospora. Proc. Nat. Acad. Sci. U. S. A. 27 (11), 499–506.

Bell, J. E. and E. T. Bell. 1988. Proteins and Enzymes. Englewood Cliffs, NJ: Prentice Hall.

Bergmann, M. and Zervas L. 1932. Über ein allgemeines Verfahren der Peptid-Synthese. *Berichte der Deutschen Chemischen Gesellschaft* 65(7), 1192–1201.

Berzelius, J. J. 1838. The word "protein" was coined from Greek word proteios meaning the first by Jöns Jakob Berzelius in 1838 in a letter to his friend.

Brenner, S., F. Jacob, and M. Meselson. 1961. An unstable intermediate carrying information from genes to ribosomes for protein synthesis. Nature 190, 576–581.

Crick, F. H. C. 1958. Biosynthesis of macromolecules. Symp. Soc. Exp. Biol. XII, 138–163.

Crick, F. H. 1970. Central dogma of molecular biology. Nature 227, 561–563.

Darwin, C. 1859. On the Origin of Species, 1st ed. London, UK: John Murray.

Dyson, H. J. and P. E. Wright. 2005. Elucidation of the protein folding landscape by NMR. Methods Enzymol. 394, 299–321.

du Vigneaud, V., C. Ressler, J. M. Swan, C. W. Roberts, and P. G. Katsoyannis. 1954. Oxytocin: synthesis. *J. Am. Chem. Soc*. 76; 3115–3118.

Edman, P. 1950. Method for determination of the amino acid sequence in peptides. Acta Chemica Scandinavia. 4, 283–284.

Edman, P. and G. Begg. 1967. A protein Sequenator Eur. J. Biochem. 1, 80–91.

Faustino, N. A. and T. A. Cooper. 2003. Pre-mRNA splicing and human disease. Genes Dev 17, 419–437.

Fire, S. Q., M. K. Xu, S. A. Montgomery, S. E. Kostas, and C. C. Driver. 1998. Potent and specific genetic interference by double-stranded RNA in Caenorhabditis elegans. Nature. 391, 806–811

Fischer, E. 1902. *Nobel Lectures, Chemistry 1901–1921*, Amsterdam, The Netherlands: Elsevier 1966

Gorrod, A. E. 1909. Inborn Errors of Metabolism. Oxford, UK: Oxford University Press.

Hershey, A. D. and M. Chase. 1952. Independent functions of viral protein and nucleic acids in growth of bacteriophage. J. Gen. Physiol. 36, 39–56

Ingram, V. M. 1956. A specific chemical difference between globins of normal and sickle-cell anúmia hemoglobins. Nature 178, 792–794.

Ingram, V. M. 1957. Gene mutations in human hemoglobin: the chemical difference between normal and sickle húmoglobin. Nature 180, 326–328.

Jacob, F., and J. Monod. 1964. Biochemical and genetic mechanisms of regulation in the bacterial cell. Bull. Soc. Chim. Biol. 46, 1499–1532.

Kendrew, J. 1961. The three-dimensional structure of a protein molecule. Sci. Am. 205, 96–110.

Khorana, H. G. 1968 Nucleic acid synthesis in the study of Genetic code. Nobel Lecture 341–366.

Klose, J. 1975. Protein mapping by combined isoelectric focusing and electrophoresis in mouse tissues. A novel approach to testing for induced point mutations in mammals. Humangenetik 26: 231–243.

Lander, E. et al. 2001. International Human Genome Sequencing Consortium: Initial sequencing and analysis of the human genome. Nature 409, 860–921.

Leder, P., and M. W. Nirenberg. 1964. RNA codewords and protein synthesis3. On the nucleotide sequence of a cysteine and leucine RNA code words. Proc. Na. Acad. Sci. U.S.A. 52, 1521–1529.

Lewin, B. 2004. Gene VIII. Upper Saddle River, NJ: Prentice Hall.

Mattick, J. S. 2003. Challanging the dogma: the hidden layer of non-protein-coding RNAs in complex organisms. BioEssays 25, 930–939.

Mendel, J. G. 1866. Versuche über Plflanzenhybriden Verhandlungen des naturforschenden Vereines in Brünn, Bd. IV für das Jahr, 1865. Abhandlungen, 3–47. For the English translation, see: Druery, C. T. and W. Bateson. 1901. Experiments in plant hybridization. J. R. Horticul. Soc. 26, 1–32.

Merrifield, R. B. 1963. Solid Phase Peptide Synthesis. I. The Synthesis of a Tetrapeptide. J. Am. Chem Sci J. 85(14), 2149–2154.

Mishra, N. C. 2002. Nucleases—Molecular Biology and Applications. New York: Wiley.

Mitchell, H. K. and J. Lein. 1948. A Neurospora mutant deficient in the enzymatic synthesis of tryptophan. J. Biol. Chem. 175, 481–482.

Mitchell, H. K., M. B. Houlahan, J. Lein. 1948. Some aspects of genetic control of tryptophan metabolism in Neurospora. Genetics 33, 620.

Moore, S. and W. Stein. 1972. The chemical synthesis of pancreatic ribonuclease and deoxyribonuclease. Nobel Lecture 80–93.

Nilsson, B. L., M. B. Soellner, and R. T. Raines. 2005. Chemical synthesis of proteins. Annu. Rev. Biophys. Biomol. Struct. 34, 91–118.

O'Farrell, P. H. 1975. High resolution two-dimensional electrophoresis of proteins. J. Biol. Chem. 250, 4007–4021.

Pauling Linus 1954 in Nobel Lectures, *Chemistry 1942-1962*, Elsevier Publishing Company, Amsterdam, 1964.

Perutz, M. F. Rossman, MG; Cullis, AF; Muirhead, H; Will, G; North, 1960. Structure of haemoglobin: a three-dimensional Fourier synthesis at 5.5A resolution, obtained by x-ray analysis. Nature 185, 416–422.

Prucca, C. G., I. Salvin, R. Quiroga, E. V. Elias, F. D. Rivero, A. Saura, P. G. Carranza, and H. D. Lujan. 2008 Antigenic variation in Giardia lamblia is regulated by RNA interference. Nature 456, 750–754.

Sanger, F. 1952. The arrangement of amino acids in proteins. Adv. Protein Chem. 7, 1–28

Sanger, F. 1958. The chemistry of insulin. Nobel Lecture 544–556.

Sarabhai, A. S., A. W. O. Stretton, S. Brenner, and A. Bolle. 1964. Colinearity of the gene with the peptide chain. Nature 201, 13–17.

Sharp, P. A. 2005. The discovery of split genes and RNA splicing. Trends Biochem Sci. 30, 279–81.

Stryer, L. 1982. Biochemistry, 2nd edition. San Francisco, CA: W.H. Freeman Co.

Sumner, J. B. 1946. The chemical nature of enzyme. Nobel Lectures 114–121

Tanford, C. and J. Reynolds. 2004. Nature's Robot: A History of Proteins. Oxford, UK: Oxford University Press.

Volkin, E. and L. Astrachan. 1957. Phosphorus incorporation in Escherichia coli ribo-nucleic acid after infection with bacteriophage T2. Virology 1956, 149–161.

Watson, J. D. and F. H. C. Crick 1953a. Molecular structure of nucleic acids: A structure for desoxyribonucleic acids. Nature 171 731

Watson, J. D. and F. H. C. Crick 1953b. General implications of the structure of desoxyribonucleic acids. Nature 171. 964

Watson, J. 1965. Molecular Biology of Gene. Melno Park, CA: W. A. Benjamin.

Wilkins, M. 1996. 1997. Protein identification in the post-genome era: the rapid rise of proteomics. Q. Rev. Biophys. 30(4), 279–331.

Yanofsky, C. 1952. The effect of gene changes on tryptophan desmolase formation. Proc. Nal. Acad. Sci. U.S.A. 38, 215–226.

Yanofsky, C., B. C. Carlton, J. R. Guest, D. R. Helinski, and U. Henning. 1964. On the colinearity of gene structure and protein structure. Proc. Nat. Acad. Sci. U.S.A. 51, 266–27

Yanofsky, C. 2005a. The favorable features of Tryptophan synthetase for proving Beadle and Tatum's one gene—one enzyme hypotheis. Genetics 169, 511–516.

Yanofsky, C. 2005b. Using studies on Tryptophan metabolism to answer basic biological questions. J. Biol. Chem. 278, 19859–10878.

FURTHER READING

Anraky, Y., R. Mizutani, and Y. Satow. 2008. Protein splicing: Its discovery and structural insight into novel chemical mechanisms. IUMBMB 57, 563–574.

Baltimore, D. 1971. RNA viruses. Bacteriol. Rev. 35, 235.

Baltimore, D. 1975. Viruses, Polymerase and Cancer. Nobel Lectures. Amsterdam, The Netherlands: Elsevier, pp. 215–226.

Beadle, G. W. 1958. Genes and chemical reactions in Neurospora. Nobel Lecture. 587–599.

Cech, T. R. 1988. Conserved sequences and structure of group I intron: Building an active site for RNA catalysis—a review. Gene 73, 259.

Chou, K. C. and Y. D. Kai 2004. A novel approach to predict active sites of enzyme molecules. Proteins 55, 77–82.

Hozumi, N. and Tonegawa, S. 1976. Evidence for rearrangement of immunoglobulin genes coding for Variable and constant regions Proc. Natl. Acad. Sci. 73, 203–207.

Hozumi, N., and S. Tonegawa. 1976. Evidence for somatic rearrangement of immunoglobulin genes coding for variable and constant regions. Proc. Natl. Acad. Sci. U.S.A. 73(10), 3628–3632.

Kaiser, A. D., and D. S. Hogness. 1960. The transformation of Escherichia coli with deoxyribonucleic acid isolated from bacteriophage lambda-dg. J. Mol. Biol. 392–415.

Keta, P., D. W. Summers, H.-Y. Ren, D. M. Cyr, and N. W. Dekhelyan. 2009. Identification of a concensus motif in substrates bound by a TypeI Hsp40. Proc. Nat. Acad. Sci. U.S.A. 106, 1–5

Margulies, E. H., et al. 2005. An initial strategy for the systematic identification of functional elements in the human genome by low-redundancy comparative sequencing. Proc. Nat. Acad. Sci. U.S.A. 102, 4795–4800.

Nguyen, H. D. and C. K. Hall. 2004. Molecular dynamics simulation of spontaneous fibril formation by random coil peptide. Proc. Natl. Acad. Sci. U.S.A. 101, 16180–16185.

Stahl, N. and S. B. Prusiner. 1991. Prions and prion proteins. FASEB J. 5, 2799–2807.

Tatum, E. L. 1958A. Case history in biological research. Nobel Lecture 2–9.

Taubes, G. 1996. Misfolding the way to disease. Science 271, 1493–1495.

Thomas, P. J., B-H Qu, and P. L. Peterson. 1995. Defective protein folding as a basis of human disease. TIBS 20, 456–459.

Temin, H. 1975. DNA provirus hypothesis. Nobel Lecture 215–245

Tonegawa, S. 1987. Somatic generation of immune diversity. Nobel Lecture 380–405

Venter, J. C., et al. 2001. The sequence of the human genome. Science 291, 1304–1351.

CHAPTER 2

PROTEOMICS – RELATION TO GENOMICS, BIOINFORMATICS

Proteomics is the understanding of all proteins and their interactions in a cell or an organism; therefore, proteomics is directly related to genomics, because genomics is the understanding of all genes in any organism and the study of how genes encode for proteins. Bioinformatics or computational biology helps in the analysis and management of all data generated during the study of the genomics and proteomics of an organism. Genomics is static because the nature of genes or the DNA sequence remains the same in all cells of an organism. In contrast, proteomics is dynamic because the protein profiles change from one cell type to another or during different stages of development. The different stages in the life of an insect point out the stark distinction between genomics and proteomics: For example, the caterpillar and the butterfly possess the same genome, but their proteomes or protein profiles are sharply different, which gives them entirely different morphology and function as if these are two distinct creatures (Figure 2.1).

In humans, about 30,000 genes are responsible for making approximately 500,000 proteins. It is indeed a challenge to determine how such a small number of genes can make so many proteins. Much of the change in protein profile is generated by the alternate splicing of transcripts and posttranslational modification of proteins.

Introduction to Proteomics: Principles and Applications, By Nawin C. Mishra

Figure 2.1: The caterpillar and butterfly exemplify the differences in the proteiomics at two different stages in the life cycle of an insect. (Courtesy of Professor Richard Vogt of the University of South Carolina.)

2.1 GENOMICS

Genomics includes the determination of the entire DNA sequence in the chromosome(s) of an organism and understanding the nature of these sequences. The entire DNA sequences of more than 500 organisms have been determined. It began with the sequence of a bacterial virus $\phi \times 174$ and progressed to the genome of humans (see Table 2.1).

It seems that at least three classes of DNA exist in higher organisms. One class of DNA encodes for proteins; these DNA sequences are transcribed and then translated into proteins that control the structure and function of an organism. The second class includes only transcribed DNAs; during

Table 2.1. Genomes of different organisms.

Organism	Year	Size	Comment
$\phi \times 174$ virus	1977	3 Kb	First bacterial virus, not a free-living organism sequenced.
Hemophilus influenze	1995	1.8 Mb	First bacterium or a free-living organism sequenced.
Saccharomyces cerevisiae	1996	12 Mb	Yeast, the first eukaryote sequenced
Caenorhabditis elegans	1998	100 Mb	Soil worm
Drosophila melanogaster	2000	180 Mb	Fruit fly
Human	2000	3000 Mb	The first draft of human genome sequence announced.
Chimpanzee	2005	3000 Mb	The closet living relative of human.
Neanderthal	2006	3000 Mb	The extinct cousin of human, more than a million base pairs sequenced.

transcription into transfer RNA (tRNA) and ribosomal RNA (rRNA), these RNAs help in the translation of other sequences into proteins. The third kind of DNA sequences either during transcription into small nuclear RNA (snRNA) and microRNA or as intact DNA, such as promoter, operator, enhancer, and silencer regions control the expression of other genes. Thus, the third kind of DNA sequences serves only as a regulatory sequence to control the expression of other DNA sequences. In addition, some DNA sequences may serve as the structural components or as different kind of codes other than those required for coding proteins. These DNA sequences are required for maintaining the integrity and function of chromosomes. For example, the nucleotide sequences in telomeres do not code for any protein but are required for the maintenance of the size or integrity of chromosomes during replication. Likewise, the DNA sequences that comprise the centromeres do not code for any protein but are required for the partitioning of daughter chromosomes during cell division. Without the proper function of the centromeres, cell division would cause unequal distribution of chromosomes to the daughter cells, which results in what is called aneuploidy. Aneuploidy in humans causes several syndromes like Down syndrome and diseases including cancer. In addition, several DNA sequences act as control regions or as structural components of the chromosome in ways that are not yet elucidated. Another 30–40% of DNA sequences are simply the leftover retroposons accumulated during the evolutionary history of organisms, including humans. Thus, the nature of a large number of DNA sequences in higher organisms is still unknown. These DNA sequences have been called "junk DNA" because of the lack of understanding of their role in the structure and function of the genome.

Theoretically, genomics started with the elucidation of the double-helical structure of DNA by Watson and Crick in 1953. This structure of DNA embodies all the attributes of biological information in genes/DNA. The fact that one strand of DNA is complementary to another in a DNA double helix provides the mechanism for the replication of the genetic information by using one strand as template for the synthesis of the opposite strand in the DNA. It also provides the mechanism for the storage and transfer of genetic information. The information is stored as the combination of three adjacent nucleotides called a triplet or a codon; these are made out of the four nucleotides (adenine, cytosine, guanine, and thymine) in the DNA. The transfer of information is accomplished from DNA into messenger RNA (mRNA) via transcription and then from mRNA into proteins via translation. Finally, the DNA structure provides means for mutation or heritable change in the genetic information through accidental error during the copying process of the DNA molecules at the time of replication.

Genomics, however, remained a dream until the discovery of enzymes such as DNA polymerases, restriction endonucleases, and ligases for the synthesis, cutting, and joining of the DNA segments. The march toward genomics was propelled by the development of certain technologies such as DNA cloning, DNA sequencing (Maxam and Gilbert 1977, Sanger Nicklen and Coulson 1977), and DNA amplification (Mullis and Faloona 1987). The automation of DNA sequencing at the large scale (Smith et al. 1985) facilitated the sequencing of more than a million nucleotide per day. The development of computers and related programs made the automation of DNA sequencing possible. The use of computers also made it feasible to manage and analyze the DNA sequence data leading to the beginning of genome projects (Collins and Galas 1993).

2.1.1 Human Genome Project and Other Genome Projects

Once the technology for sequencing DNA pieces was available, it became imperative to develop projects to unravel the DNA sequence of humans and other organisms to understand fully how the DNA carries the information for the development of different organisms into adult form, which includes a single fertilized cell and their interrelationship. To achieve this goal, attempts were made to develop and improve the technology for sequencing the DNA by throughput and automation and then to undertake the genome projects. Both aspects of DNA sequencing progressed side by side. In 1990, it was proposed by the U.S. Department of Energy and by the National Institutes of Health to undertake the Human Genome project to determine the entire DNA sequence of humans to understand its role in human development and health. The genome projects of several genetically defined organisms were also undertaken as model systems for understanding the role of genes that could not be studied in humans for technical and ethical reasons. In June 2000, the draft copy of the human DNA sequence was announced and published, which was christened by W. Clinton, then the President of the United States, as the language of God in which our fate is written. The entire genomes of several model organisms were determined before the completion of the human genome project. These included Hemophilus, yeast, fruit fly *Drosophila melanogaster*, roundworm *Caenorhabditis elegans*, and *Arabdopsis thaliana*, a member of the mustard family. The DNA sequence of Hemophilus showed that as little as 460 protein-encoding genes are enough to sustain the life of a unicellular organism. The DNA sequence of yeast showed the minimal requirement of approximately 6000 genes for an eukaryotic cell. The analysis of the yeast genome also revealed the proportion of genes required for different cellular metabolic pathways. The DNA sequence of the fruit fly showed the presence of approximately 13,000 genes required for the complete development of a complex multicellular organism. The genome study of the roundworm showed the existence of about 19,000 genes. This study also provided the role of programmed cell death in the development of this worm. The most surprising fact that emerged from the study of the human genome was the meager number of about 23,000 genes required for the complete development of a complex organism from a single fertilized cell into an adult containing 10^{13} cells with more than 700 cell types organized into different tissues and organs. Initially, it was thought that humans possesses as many as 100,000 genes. This estimate was based on the hypothesis of one gene for every protein, but it turns out that the human genome contains less than 23,000 genes and still possesses more than 100,000 proteins made possibly by alternate splicing of the transcripts from a small number of genes. The human genome project not only determined the entire sequences of about 3 billion nucleotides in human distributed over 24 chromosomes including 22 autosomes and X and Y chromosomes but also made the technology developed during this process available to the public. This project, for the first time, tried to understand the ethical, legal, and social implications (ELSI) of the findings. This effort marked a great departure from the scientific efforts of the past. The sensitivity toward developing ethical questions was unlike any other efforts in the past. For example, when the Manhattan Project to develop a nuclear device was undertaken, the ethical concerns were totally ignored, which atomic scientists regretted later.

2.1.2 Methods to Study the Genome Project

The development of recombinant DNA technology and the discovery and construction of vectors were instrumental in molecular cloning. Human DNA or other DNA could be readily obtained as fragments of desired lengths after treatment with restriction endonucleases. These fragments then could be inserted with the help of ligases into circular vector DNA, such as plasmid DNA, which has been linearized by a cut with the restriction enzyme at a unique site.

The recombinant vector DNA containing the human or other alien DNA segments as inserts could be introduced into bacterial host cells such as *Eserichia coli* via the process of transfection (see Figure 2.2). The introduced recombinant vectors are propagated in bacterial host cells because the vectors are capable of self-replication in the host cells. Many methods are available to amplify the number of the chimerical vectors greatly in the host cells. In addition to plasmid DNA, many other kinds of vectors are now available. These include cosmid, λ phage DNA, YAC (yeast artificial

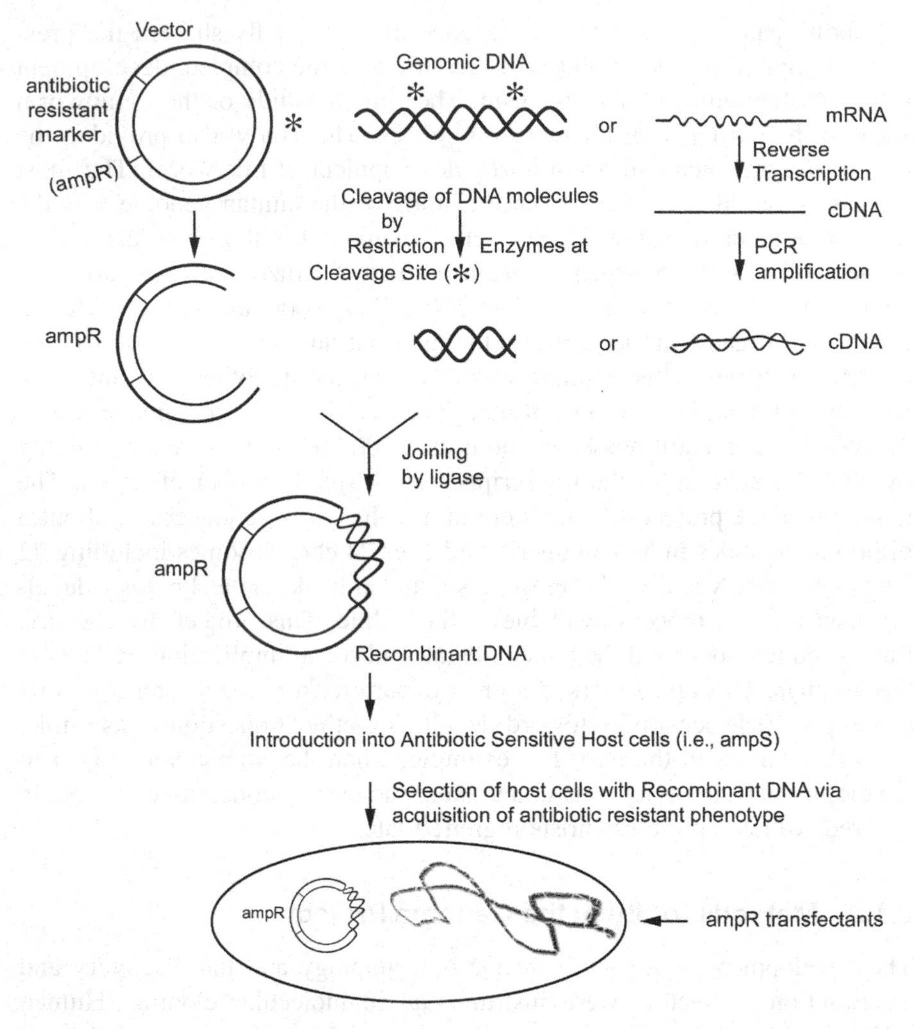

Figure 2.2: A general method for the cloning of a gene. (Reproduced from Mishra, 2002, with the permission of John Wiley & Sons.)

chromosome), BAC (bacterial artificial chromosome), and PAC (phage artificial chromosome). These vectors are useful for the cloning of large pieces of DNA segments and, thus, are also useful in the genome project. Before the start of the human genome project, a several human genes of interest, such as those that are responsible for causing human diseases, were cloned by the newly developed methodology of DNA cloning as described here. In this approach to clone human genes or genes of interest from other sources, random DNA segments were first cloned and then bacterial cells harboring the cloned DNA segments of interest were selected. This method of cloning was called the shotgun method. Usually, a collection of bacterial colonies containing the entire genome of an organism is made. Such a collection is called the genomic library. In addition to the genomic library obtained by the shotgun method, there are other kinds of DNA libraries. The cloned DNA segments from any of these libraries are then used for the purpose of sequencing. At times, these DNA segments are amplified in vitro by the method of polymerase chain reaction (PCR), which was developed by Mullis in (1985) (see Mullis & Faloona 1987). Mullis was awarded the Nobel Prize for developing the PCR methodology. The DNA sequences thus obtained are arranged by the computers into the nucleotide sequence of a chromosome.

Subsequent to the development of a genomic library, more specific methods were developed to clone DNA from a particular chromosome or from a specific tissue expressing a particular gene. Thus, different kinds of DNA libraries became available. These are the genomic library, chromosome library, and expression library. A collection of bacterial cells harboring the entire genome of an organism obtained by the shotgun method of cloning is called a genomic library. The bacterial cells containing the collection of DNA segments from a particular chromosome are called a chromosome library. In contrast, the expression library contains a collection of expressed genes/DNA segments from a particular tissue such as blood cells, brain cells, or liver cells. The expression library is constructed from the mRNA of a particular cell type. The mRNAs are isolated from a particular tissue and then reverse transcribed into complementary DNA (cDNA) with the help of the enzymes reverse transcriptase. The single-stranded cDNA thus obtained is made double stranded with the use of DNA polymerase and then inserted into vector DNA for propagation in bacterial host cells during the construction of the expression library.

The genomic library and the chromosome library contain clones of both the protein coding as well as the protein noncoding segments of DNA from an organism. However, it requires a significant amount of effort to place them into chromosomes for the purpose of mapping and generating the entire DNA sequence of a chromosome. The chromosome-specific library is somewhat easier for the purpose of mapping. It provides the DNA sequence of a chromosome directly from one end to another end. It is useful to generate overlapping DNA segments from a chromosome for mapping. With the help of overlapping regions, it is possible to walk from one end of the chromosome to the other end.

The chromosomes of an organism must be separated from each other and obtained in pure forms to construct the chromosome library. The number and size of chromosomes vary greatly in different organisms. The chromosomes of lower eukaryotes such as yeasts are much smaller in size, because they contain about 15 million base pairs or less. They are separated by pulse field gel electrophoresis. The chromosomes of humans or other higher organisms are much larger in size. Human chromosomes are 150 million base pairs long on average. Human chromosome 1 contains more than 260 million base pairs, whereas the Y chromosome has about 60 million base pairs. They could not be separated until the cell sorter was developed at the Los Alamos National Laboratory in the late 1980s.

2.1.3 Outcome of the Study of the Human Genome

2.1.3.1 Low Number of Human Genes and Gene Discovery. One of the biggest surprises of the human genome project was the low number of protein-encoding genes. It was thought initially that humans may have as many as 100,000 genes because it was thought that humans may have about 100,000 functions, each of which is controlled by one enzyme. This view was consistent with the one-gene–one-enzyme concept of Beadle and Tatum (1941) and persisted when the mRNA array for the analysis of mRNA became available. An array analysis of the mRNA showed the presence of approximately 100,000 mRNAs. However, when the nucleotide sequence for humans became available after the completion of the human genome project and a search for the protein encoding genes by virtue of their possession of nucleotide sequences characteristics of the starting and end points in a gene were looked for, it became obvious that only about 23,000 protein encoding genes exist in humans.

A large number of mRNAs and proteins in human was generated from a small number of genes through alternate splicing of the genes. It is now estimated that more than 50% of human genes undergo alternate splicing, which produces about three proteins per gene on average. On the surface, it seems amazing that humans (23,000 genes) have only a few thousand genes more than the number of genes possessed by a fruit fly (13,000 genes) or a roundworm (19,000 genes). Still, it is conceivable that just a few thousand genes could make such a huge difference and could lead to the development of an organism as complex as a human. This view becomes comprehensible when it is recognized that humans and chimpanzees differ by about a few hundred genes; of that, less than 10 genes can account for anatomical differences, and a maximum of 40 genes can account for the differences in cognitive development.

2.1.3.2 Junk DNA. The human genome project revealed the presence of DNA sequences that do not code for any protein. These nonprotein coding sequences amount to about 95% of the total human nucleotide sequences. These DNA sequences do not constitute typical genes. Because they do not directly control human traits, they have been called junk DNA. However, the so-called junk DNA are not junk at all, because they have been retained in the evolutionary history of mankind and a mutation in them may stop human development.

These sequences are junk DNA in the same way we save the junk in our attic with the anticipation that we may need it someday. They are certainly not garbage DNA; if this were true, these DNA would have been eliminated during the evolutionary process, the same way we throw away our garbage. Even before the human genome project, there were solid indications for the existence of the so-called repeat sequences in higher organisms, including humans, from the DNA hybridization experiments by Britten and Kohne (1968).

The human genome project has confirmed the existence of these repeat or junk DNA sequences and has established their location on the chromosomes. More than 50% of these repeat sequences consist of repeat bases such as GCGCGC. These have been duplicated as multiple copies and have been distributed randomly over the genome in between the protein coding sequences. In DNA hybridization experiments, these sequences with multiple copies appear as a rapidly renaturing fragment of the genome. In contrast, the protein coding sequences appear as a slowly renaturing fragment of the genome, because they are present as a single copy of the DNA sequence in the genome. Many of these sequences reveal our evolutionary past. Among the junk DNA is a short sequence of about 300 base pairs called Alu sequences. Alu sequences constitute about 7% of the human genome (i.e., much more than protein coding sequences, which represent only 5% of the genome). The so-called junk DNA including the Alu sequences has been called "selfish DNA." They are selfish in the sense that their sole purpose is to perpetuate as a part of the genome. These sequences could not be eliminated during the evolutionary process. The Alu sequences are characteristic of the primates.

It is a great challenge to understand their role in humans and other higher organisms. It is assumed that some of the junk DNAs are only the RNA transcribing sequences that have roles in the process of translation or in the control of gene action or gene silencing. Other junk DNAs may play a variety of structural roles in connecting the genes, determining the spacing between the genes, or controlling the supercoiling of a chromosome and overall integrity of the chromosome.

2.1.3.3 Restriction Fragment Length Polymorphism (RFLP). This polymorphism consists of differences in the length of a particular DNA segment obtained after digestion with a restriction enzyme. For example, normal and sickle cell persons may have different lengths of

DNA segments that contain the hemoglobin gene. This was first discovered by Kan and Dozy (1978). They digested the DNA of a healthy person and that of a sickle cell person with a particular restriction enzyme and then identified the DNA segments that contain the hemoglobin gene after Southern hybridization (Southern 1975) using the cloned P32-labeled hemoglobin gene as a probe. They found that the hemoglobin gene containing the DNA segment was cut into two fragments in a healthy person, but this gene appeared as a much larger single-fragment gene in a sickle cell person. In the sickle cell person, the presence of a single fragment was a result of the loss of an internal restriction site. Such polymorphisms were called RFLPs. RFLPs may develop because of mutation that led to the loss or gain of a restriction site in a particular DNA segment. This polymorphism may also develop based on variations in the number of certain terminal repeats (VNTRs) present in a DNA segment. The RFLP was used for mapping the human chromosome and in the identification and location of a disease-causing gene on a chromosome (Botstein et al. 1980). Thus, RFLPs became an essential prelude to the human genome project.

2.1.3.4 Single-Nucleotide Polymorphism (SNP). Before the genome project, it was assumed that all normal human individuals possessed identical nucleotide sequences in their chromosomes. It was also considered that those who suffered from a disease differed from the normal individuals only in the DNA sequence of the gene(s) involved in causing the disease. However, once the entire genome sequences of many normal individuals were deciphered, it became obvious that they differed in their nucleotide sequences in one or more places. This difference in the nucleotide sequences among the normal individuals was called an SNP. Thus, it is possible that one individual may possess a sequence of AAGCCTA in a particular gene, whereas another individual may have AAGCTTA in the same sequence within that gene showing an SNP. Thus, these two individuals represent two separate alleles of that gene (i.e., the C allele and the T allele). A population in which 1% or more individuals differ in their DNA sequence is ordinarily considered as having an SNP. It is important to mention that SNPs are characteristic of a population that belongs to an ethnic group or geographical location. SNPs may occur in the coding or noncoding sequences of a gene or in the intergenic regions between genes. SNPs in the coding region of a gene may still code for the same amino acid in the protein because of degeneracy of the genetic code. SNPs that lead to formation of the same protein are called synonymous, whereas the SNPs that produce different proteins are called nonsynonymous. SNPs in the noncoding or intergenic regions may cause defects in splicing or in transcription factor binding, or SNPs may cause changes in the nature of the noncoding or regulatory RNAs.

SNPs occur throughout the entire human genome. They occur with a frequency of 1 in every 300 nucleotide stretches of DNA. There are roughly 10 million SNPs in the human genome containing 3 billion nucleotides. In Two thirds of all SNPs, cytosine is replaced by thymine. SNPs are usually detected by their RFLP because of changes in the nucleotide sequence identified by a restriction enzyme. SNPs are more readily detected by microarray analysis or by DNA sequencing. SNPs are considered valuable in developing personalized medicine because SNPs form the basis of an individual response to a drug, chemical, or pathogen.

2.1.3.5 Copy Number Variation (CNV). Chromosomes of different individuals carry the same number of genes or DNA sequences.

A loss of a segment in a chromosome is called deletion, and a gain in the number of a particular segment is called duplication; these are the main causes of several human diseases or syndromes. However, the results of the human genome project have revealed the presence of a different number of the same DNA segments with or without any disastrous effect on the individual. Thus, a variation in the copy number of a stretch of DNA in the human chromosome is another important finding from the human genome project. It has been found that a stretch of DNA may exist in more than one copy in the homologous chromosome of the same individual or in different individuals with or without causing any apparent health problems. This has been called CNV. The variation caused by different copy numbers of the DNA segment far exceeds the variation caused by SNPs.

CNV has been found to affect the expression of genes or the resistance to HIV and malaria. CNV may cause certain immunological disorders or neoplasia and may also be the basis for complex diseases such as diabetes and heart disease. Furthermore, CNV may affect the adaptability of an individual in a particular environment. In the human genome, more than 1400 CNVs have been reported. There are many hot spots for CNVs in human and Chimpanzee chromosomes. Certain hot spots are common to both organisms, which suggests their evolutionary significance. Most amazingly, 80% of identical twins have been found to possess CNVs for certain DNA segments. Such CNVs may be the basis for phenotypic differences among identical twins. Recently, CNV has been found to occur in the DNA of patients with neuroblastoma, autism, and Alzheimer diseases.

2.1.3.6 Primate Specific Genes. A recent analysis of human complemantary DNA (cDNA) and expressed sequence tag has led to identification of over 38 thousand human transcriptional units. A comparison of these human transcriptional units with other primates and non-primate sequences has identified 131 primate specific transcriptional units that are found in primates only. Among these transcriptional units shared by the primates about half of them possess protein–coding sequences. These were also found to lack introns indicating that they might have originated from transposons. Study of the expression pattern of these primate specific genes suggest these transcripts to be expressed only in brain and reproductive tissues. About 21 of these primate specific genes were found to be involved in the development of human sperms and many causing infertility like tetrazoospermia in human. It is concluded that these primate specific genes influenced the evolution of primates.

2.1.3.7 Uniqueness of Human. Humans are distinguished from other animals and from other primates, particularly chimpanzees, our closest living relatives, in many ways. This includes our ability for bipedal, straight, upright walking and our ability for speech, language development, and many other sophisticated features that are the basis of our culture and civilization, which include agriculture, architecture, gourmet cooking, music, painting, advanced weaponry systems, the ability to discover and invent things, and a passion to comprehend the principles of nature. So what makes us human? This is an age-old question. It seems that recent advances in molecular biology and genomics, including the comparison of human and Chimpanzee genomes, seems to provide some clue to this question.

The study of genomics suggests that rapid changes in primate genes are involved in major pathways controlling the perception of sound, nerve signal transmission, cellular ions transport, and sperm production. The rapid changes in these groups of genes set primates apart from mammals and other organisms. Certain major changes that distinguish humans from chimps involve the genes responsible for the development of bipedal walking, an enlarged brain, speech, and complex language skills. The genes responsible for these functions, except for acquisition of the ability for speech, remain to be defined at the molecular level. It seems certain that a gene for a transcription factor protein called FOXP2 on chromosome 7 is responsible for speech or language in humans. This has been elucidated by Svante Paabo and his group at the Max Planck Institute in Leipzig, Germany, by comparing the amino acid sequence of the Human FOXP2 with other organisms (Enard et al. 2002). This conclusion is supported by the finding that a British family with a severe inherited speech deficit possessed an altered form of the FOXP2 gene (Lai et al. 2001, 2003). The human FOXP2 is different from the FOXP2 genes from chimpanzees and other primates (Enard et al. 2002). The mutant human FOXP2 gene found in the British family with the speech deficit is like the chimpanzee's FOXP2.

An important fact has emerged from the study of the Neanderthal genome project: They possessed the same nucleotide sequence in the FOXP2 gene as found in humans. This finding suggests that Neanderthals had acquired the ability for speech. It is estimated that FOXP2 gene, as found in humans and Neanderthals, developed approximately 400,000 years ago in a common ancestor. Recent experiments show that transgenic mice with a human copy of the FOXP2 gene do sqeak differently, which suggests the role of the FOXP2 gene in controlling speech. Studies of these transgenic mice also suggest the role of other genes beside the FOXP2 gene in controlling speech in humans.

A family in Turkey has been found to suffer from the loss of the inherited form of upright walking ability. However, a rare mutation in the gene leading to a loss of upright walking ability in humans has been now identified. This gene controls a protein responsible for the development of the cerebellum in the brain. In addition to controlling the upright gait, this gene mutation may also cause impaired speech and mental ability. A study of similar patients in Iraq and Brazil suggests the involvement of other genes as well in controlling the ability for bipedal upright walk in humans.

In addition, humans have a much larger brain. The gene ASPM that affects human brain size has been described. Furthermore, genes HAR1 and HAR2 have been now described, which are important in determining the uniqueness of humans. These genes are called HAR (human accelarated region) because they are located in the region with high activity of mutations; for example, the HAR-1 DNA shows only one mutation difference in chimp from mouse, whereas the stretch of this DNA shows 32 mutations in humans from chimpanzees. There are at least two such genes HAR-1 and HAR-2 that are described from humans. HAR-1 codes a transcription factor that is responsible for the development of a six-layer structure of the human cortex. HAR-2 controls the development of the hands in humans. In addition to these differences, humans are unique in possessing several forms of the amylase gene and also are unique in possessing a form of the lactose gene that allows the adult human to metabolize lactose, the so-called "milk sugar". These genes allow a nutritional advantage to humans.

Both human and chimpanzee genomes contain 3 billion base pairs and are upto 99% identical. There is a difference of 1.23%, which amounts to 35 million DNA base pairs. In addition, there are about 5 million deletions and insertions of DNA segments in the human genome. Thus, there is about a 4% difference in the genome between humans and chimpanzees. Of the 25,000 genes, about 580 genes have undergone rapid and positive selection, as found with the FOXP2 gene. These genes also include HARs. Many of these genes lie in DNA sequences with little or no function. Also some of these genes are missing from the chimpanzee genome. For example, three key genes that control inflammation are missing from the chimpanzee genome, which explains the differences in the immune and inflammatory response in humans and chimpanzees. Likewise, many genes are missing from the human genome. This includes the caspase-12 gene; the presence of this gene provides protection against Alzheimer disease in chimpanzees, but its absence in humans causes Alzheimer disease. Some of these distinctive DNA segments are presented graphically in Figure 2.3. One of the major conclusions of the comparative study of human and chimpanzee genomes is that it does not take a tremendous amount of genomic changes to evolve a new species; rather, a minimum of changes can evolve a new species (Pollard 2009) just as in humans and chimpanzees.

2.1.3.8 Genomic Rearrangement. One of the major findings of genomics includes the rearrangement of the genome in different organisms during the process of development. Such genomic rearrangements have been seen in hagfish, lamprey, fly, roundworm, and many other organisms, including lower vertebrates. The genomics of the lamprey and hagfish reveal a significant reduction in the genome size of the adult cells, such as blood cells, muscle, fin tail, kidney, and liver cells as compared with sperm cells. The results of studies in the lamprey have revealed a reduction of more than 20% of the DNA in the adult cells in comparison with the DNA content of the sperm cells. It has been shown that certain DNA sequences dubbed as germ-1 and some other sequences are excluded at the time of adult development in lampreys. Understanding of the mechanism including the regulation of such rearrangement may provide a clue to the rearrangement accompanying the onset of cancer in many cell lines. This may provide a means to control these dysfunctional genomic rearrangements that lead to human disesases.

2.1.3.9 Epigenomics. Epigenomics is the genome-wide study of epigenetics. Epigenetics involves the changes in DNA sequences that influence the expression of genes and is thus directly related to proteomics. The most common change involves the modification of cytosine residues by methylation at particular sites in the DNA sequence of a gene. The methylation of cytosine in DNA causes changes in the conformation of nucleosomes by interfering with the ends of histone tails in the nucleosomes and thus leads into nonexpression of that gene. It has been shown that there are two classes of genes/DNA in the honey bee, *Apis mellifera* (Elango et al. 2009), that control differential functions. One class of genes is low in CpG dinucleotides sites and another is high in CpG content. The ones with high CpG sites are usually hypomethlated in germlines and are associated with

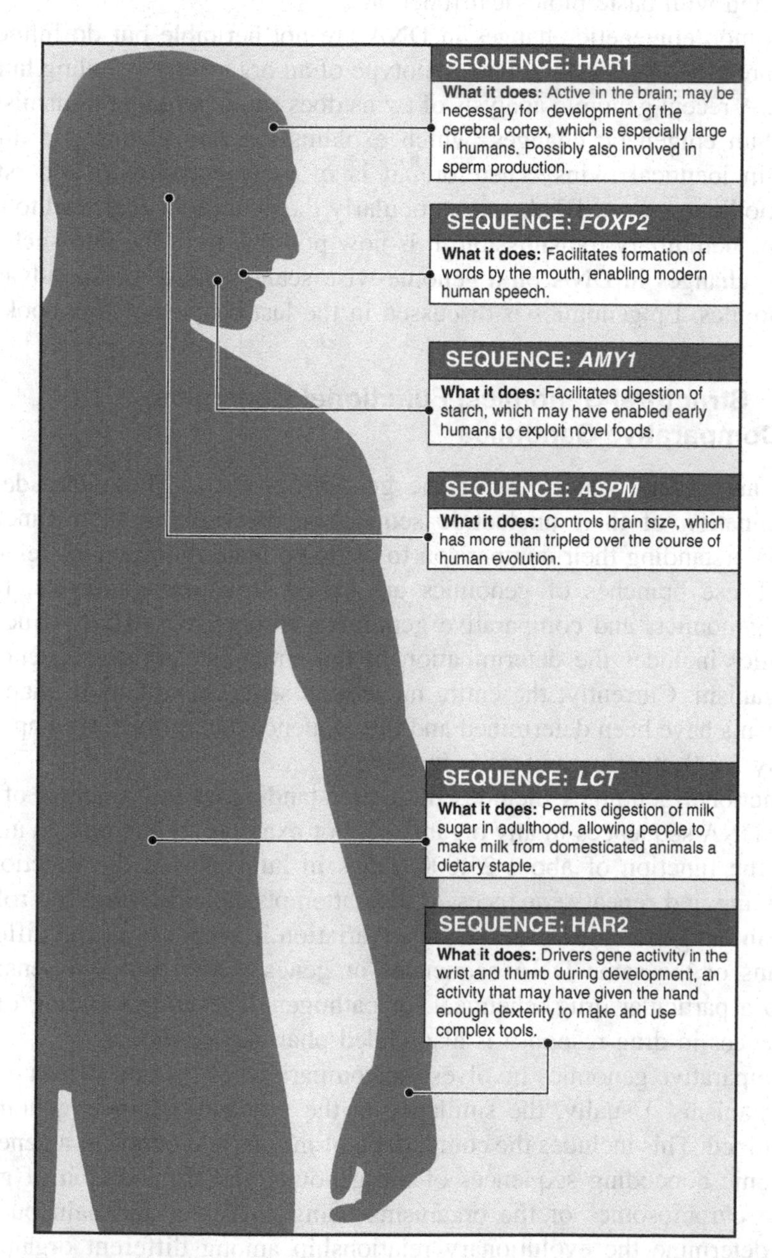

SEQUENCE: HAR1
What it does: Active in the brain; may be necessary for development of the cerebral cortex, which is especially large in humans. Possibly also involved in sperm production.
SEQUENCE: FOXP2
What it does: Facilitates formation of words by the mouth, enabling modern human speech.
SEQUENCE: AMY1
What it does: Facilitates digestion of starch, which may have enabled early humans to exploit novel foods.
SEQUENCE: ASPM
What it does: Controls brain size, which has more than tripled over the course of human evolution.
SEQUENCE: LCT
What it does: Permits digestion of milk sugar in adulthood, allowing people to make milk from domesticated animals a dietary staple.
SEQUENCE: HAR2
What it does: Drivers gene activity in the wrist and thumb during development, an activity that may have given the hand enough dexterity to make and use complex tools.

Figure 2.3: The human-specific DNA segments. (Reproduced from Pollard, 2009 with the permission of *Scientific American*.)

developmental functions, whereas the hypermethylated, low CpG genes are associated with basic biological functions.

Common epigenetic changes in DNA are not heritable but do influence the expression of genes and the phenotype of an organism, including human health. A recent genomic analysis of twins does show heritable transmission of certain epigenetic changes, which explains the rare phenotypic differences in identical twins. Therefore, it is of great concern to understand such modifications of DNA and particularly the conditions that lead to their possible hereditiary transmission. It is now possible to probe into such epigenetic changes in DNA on a genome-wise scale because of the advances in genomics. Epigenomics is discussed in the last chapter of this book.

2.1.4 Structural Genomics, Functional Genomics, and Comparative Genomics

There are several objectives of the genome projects. These include the determination of their nucleotide sequences, determining their function, and understanding their comparison to evaluate their evolutionary relationship. These branches of genomics are called structural genomics, functional genomics, and comparative genomics, respectively. Thus, structural genomics includes the determination of the entire nucleotide sequence of an organism. Currently, the entire nucleotide sequences of more than 500 organisms have been determined and the sequence determination of approximately 3000 other organisms is underway.

Functional genomics includes the understanding of the function of different DNA sequences in any organisms. For example, it attempts to understand the function of about 25,000 genes in humans and the function of noncoding and repeat sequences. It also attempts to understand the role of SNPs in the genome by observing the variation in response of the different versions of the same DNA sequences or genes to the different sensitivities to a particular drug, chemical, or pathogen. The understanding of the differences in drug response is also called pharmacogenomics.

Comparative genomics involves the comparison of genomes from different organisms. Usually, the similarity in the sequence of their genome is determined. This includes the comparison of introns and exons of a gene, the intergenic noncoding sequences of the genome, and the location of genes on the chromosomes of the organisms being analyzed. Its main purpose is to determine the evolutionary relationship among different organisms. The sequence similarity is determined by a software program called Basic Local Alignment Search Tool (BLAST). This software program is used to determine the regions of local similarity by comparing nucleotide or protein sequences with a sequence database, along with a calculation of the statistical significance of matches. BLAST is also used to assign a particular sequence to a gene family by determining both functional and evolutionary relationships between sequences.

2.1.5 Technical Advantages of Genomics

2.1.5.1 High-Throughput Methodologies. The successful completion of the human genome project and several other genome projects was made possible because of the development of high-throughput methodologies. These methods include mass-scale sequencing methodology, DNA/RNA microarray, protein chips, and mass spectrometry. These methods usually generate a tremendous amount of data within a short time.

DNA sequencing at a large scale was facilitated by the development of several high-throughput technologies using robotics and computerization. One such methodology was developed by 454 Life Sciences (Branford, CT). This methodology involves the use of picoliter reactors with robotic fluidic and optic systems for the mixing of sequencing reagents and for the detection of the product of the sequencing reaction (Margulies et al. 2005). This system is ultra fast and sequences approximately 20 million bases in about 4.5 hours. This method generates a random library of DNA fragments from the entire genome and isolates single DNA molecules on a bead, amplifies them by PCR, and sequences them. This method does not require subcloning of DNA fragments in bacteria or the handling of individual clones. This method essentially uses a pyrosequencing on a solid support in picoliter volume. Pyrosequencing involves sequencing by synthesis. In this method, a template DNA segment is copied, and every time a nucleotide is added to the growing chain, a series of enzymatic reactions triggers a chemoluminoscence or a light signal that is captured by the charge-coupled device (CCD) camera, which is analyzed by a computer program to generate the nucleotide sequence. This technology represents a tremendous advance over all technologies using Sanger's method (Sanger et al. 1977). It is of interest to mention that VisiGen Biotechnologies, Inc. (Houston, TX) is developing a DNA sequencing machine that will sequence 1 million bases per second. This facility will be available by 2010. This machine is also based on the principle of sequencing by synthesis like other throughput methods. This will sequence a human-sized genome of 3.3 billion bases in 1 hour. This instrument will be of great use in clinical application and for personalized medicine.

DNA microarray is another throughput method that determines the expression of a large number of genes under various conditions in one shot. In this method, first the DNA sequences representing all or most genes of an organism are placed at a predetermined position on a glass chip, which is later hybridized to cDNA of mRNA from a cell. The cDNA prepared out of mRNA from two different cell lines or from a cell line grown under two different growth conditions are color tagged by binding to two different fluorophores such as rhodamine (red) and fluorescein (green). Hybridization of colored cDNA with DNA sequence on the glass chip can process colors of different kinds. For example, cDNA from mRNA derived from yeast cells grown under aerobic and anaerobic conditions will show only a red color for the genes that are expressed under aerobic conditions but it will show green colors for the genes expressed under anaerobic conditions. It will show yellow colors for the genes being expressed under both aerobic and anaerobic conditions. Likewise, a normal cell may produce red colors and a cancer cell may produce green colors for the genes expressed exclusively by these two cell lines. However, the genes expressed by the two cell lines will be visualized as yellow colors. This method was first developed by Schena and others at Stanford University in 1995 (Schena et al. 1995) and has been advanced in many ways since then. DNA array for gene expression is based on DNA–DNA hybridization as done during Southern hybridization (Southern 1975). A chip containing the entire genes of an organism, such as 6000 genes of yeast, can be prepared readily by the methodology available currently. In such DNA array analysis, the intensity of color may be used to estimate the expression level of a particular gene.

Protein array has been developed to determine the expression of protein profiles of a cell. This is based on protein–protein interactions, such as antigen and antibody interactions used in a Western hybridization. In this method, a large number of antibodies or other ligands is put on specific positions on a glass chip and then analyzed by interaction with rhodamine or fluorescein-labeled protein lysates from two cell lines or from a cell line grown under two different conditions. The results are visualized as red, green, and yellow spots. Here again, the intensity of a color may be used to indicate the level of protein expression. Protein chips are of great use in proteomics for the analysis of protein–protein interactions, protein modifications, or even for the identification of substrate for enzymes. For some of these analyses, the protein chip is prepared by placing glutathion-fused ORF products of a cell. This method is of great help to drug discovery or for detection of proteins involved in the control of a disease.

Mass spectrometry is the last in the series of high-throughput methods that has sped up the identification of proteins by determining the sequence of the peptides and comparing the latter with the proteins encoded by the DNA sequence in the genome data bank. The entire process is accomplished within a few minutes using protein available in nanoscale. A single protein isolate or the entire collection of proteins from a cell is digested enzymatically, and then the mass of the resulting peptide is determined by mass/charge ratio in a mass spectroscope and matched with the protein sequence in the protein data bank to determine the identification of single or collection of proteins.

2.2 BIOINFORMATICS AND COMPUTATIONAL BIOLOGY

The term "bioinformatics" was coined by Paulien Hogeweg (1978) of Utrecht University in the Netherlands to designate the branch of science using informatics in biological systems. Commonly, bioinformatics and computational biology are used as interchangeable terms. The main objectives of bioinformatics are to generate and manage the vast amount of data in biology with the help of computers and to retrieve these data and their analysis to provide models for checking the validity of the major hypotheses or principles underlying biology. This involves the use of various disciplines, including mathematics, statistics, informatics, and artificial intelligence to solve the problems in modern biology at the molecular level. The integration of thousands of small DNA sequences into the DNA sequence of the entire genome was made possible by the application of bioinformatics. Completion of the genome project would not have been possible without advances in computational biology. The assembly of a large number of DNA sequences into the human genome is certainly the greatest triumph of bioinformatics.

Computational biology involves finding genes and determining their assembly in the chromosomes. Such analysis includes the annotation of DNA sequences to identify the genes. Marking of a DNA sequence with genes is called "annotation." Bioinformatics also includes alignments of nucleotide sequences in the genomes and of amino acids sequences in the proteins to predict the structure, function, and evolution of genes and proteins. Most of the time, homology modeling is used to predict the function of a protein by comparing the amino acid sequence of this protein with an unknown function with that of a protein with a known function. In such comparison, a protein of unknown function is inferred to have the function of the protein with a known function if the amino acid sequence of the protein in question bears similarity with the amino acid sequence of the protein whose function is already established. Sometimes, two proteins (despite the differences in their amino acid sequences) may show similarity in their protein structure and therefore are assumed to have the same function. This situation is exemplified by the human hemoglobin and plant leghemoglobin. These two proteins possess completely different amino acid sequences, but their protein structure is almost identical. Both of these proteins serve the function of carrying oxygen in the two different organisms based on a homology in the overall structures of these two proteins, despite differences in their amino acid sequence.

A large number of software programs has developed over the years; an online list of these software programs is available from Steven L. Saltzberg at the University of Maryland Center for Bioinformatics and Computational Biology (www.cbcb.umd.edu) and at The Institute for Genomic Research, Johns Hopkins University. These computational programs use the techniques of data mining, algorithms, artificial intelligence, and many more. These techniques are used for genome assembly, gene annotation, DNA and protein sequence alignments, and prediction of structure and function of gene and protein and protein–protein interactions.

REFERENCES

Beadle, G. W. and E. L. Tatum, 1941. Genetic control of biochemical reactions in Neurosporda. Proc. Natl. Acad. Sci. U.S.A. 27, 499–506.

Botstein, D., R. L. White, M. Skolnick, and R. W. Davis. 1980. Construction of genetic linkage in man using restriction fragment length polymorphism. Am. J. Hum. Genet. 32, 314.

Britten, R. J., and D. E. Kohne. 1968. Repeated sequences in DNA. Hundreds of thousands of copies of DNA sequences have been incorporated into the genomes of higher organisms. Science 161, 529–540.

Collins, F., and D. Galas. 1993. A new five-year plan for the U.S. human genome project. Science 262, 43–46.

Elango, N., B. G. Hunt, M. A. D. Goodisman, and S. V. Yip. 2009. DNA methylation is widespread and associated with differential gene expression in castes of the honeybee, Apis mellifera. Proc. Natl. Acad. Sci. U.S.A. (in press).

Enard, W., M. Przeworski, S. E. Fisher, C. S. L. Lai, V. Wiebe, T. Kitano, A. P. Monaco, and S. Paabo. 2002. Molecular evolution of FOXP2, a gene involved in speech and language. Nature 418, 869–872.

Hogeweg, P. 1978. Simulating the growth of cellular forms. Simulation 31, 90–96.

Hogeweg, P. and B. Hesper. 1978. Interactive instruction on population interactions. Comput Biol Med 8, 319–327.

Kan, Y. W. and A. M. Dozy. 1978. Polymorphism of DNA sequence adjacent to human beta-globin gene: Relationship to sickle mutation. Proc. Natl. Acad. Sci. U.S.A. 75, 5637.

Lai C. S., D. Gerrelli, A. P. Monaco, S. E. Fisher, A. J. Copp. 2003. FOXP2 expression during brain development coincides with adult sites of pathology in a severe speech and language disorder. Brain 126, 2455–2462.

Lai, C. S., S. E. Fisher, J. A. Hurst, F. Vargha-Khadem, A. P. Monaco. 2001. A forkhead-domain gene is mutated in a severe speech and language disorder. Nature 413, 519–523.

Margulies, M., M. Egholm, W. E. Altman, S. Attiya, J. S. Bader, L. A. Bemben, J. Berka, M. S. Braverman, Y. J. Chen, Z. Chen et al.. 2005. Genome sequencing in microfabricated high-density picolitre reactors. Nature 437, 376–380.

Maxam, A. M. and W. Gilbert. 1977. A new method for sequencing DNA. Proc. Natl. Acad. Sci. U.S.A. 74, 560.

Mullis, K. B. and F. A. Faloona. 1987. Specific synthesis of DNA in vitro via a polymerase catalyzed chain reaction. Methods Enzymol. 155, 335.

Pollard, K. S. 2009. What makes us human? Sci. Am. 300, 44–49.

Sanger. F., S. Nicklen, and A. R. Coulson. 1977. DNA sequencing with chain-terminating inhibitors. Proc. Nat. Acad. Sci. U.S.A. 74, 5463–5467.

Schena, M., D. Shalon, R. W. Davis, and P. O. Brown. 1995. Quantitative monitoring of gene expression patterns with a complementary dna microarray. Science 270, 567–570.

Smith, L. M., Fung, S., Hunkapiller, M. W., Hunkapiller, T. J., and Hood, L. E. 1985. The synthesis of oligonucleotides containing an aliphatic amino group at the 5′ terminus: Synthesis of fluorescent DNA primersfor use in DNA sequence analysis. Nucleic Acids Res. 13, 2399–2412.

Southern, E. M. 1975. Detection of specific sequences of DNA fragments separated by gel electrophoresis. J Mol. Biol. 98, 503.

FURTHER READING

Barnes, M. R. and I. C. Gray. 2003. Bioinformatics for Geneticists, 1st ed. New York: Wiley.

Baxevanis, A. D., G. A. Petsko, L. D. Stein, and G. D. Stomo. 2007. Current Protocols in Bioinformatics. New York: Wiley.

Claverie, J. M. and C. Notredame. 2003. Bioinformatics for Dummies. New York: Wiley.

Durbin, R., S. Eddy, A. Krogh, and G. Mitchison. 1998. Biological Sequence Analysis. Cambridge, UK: Cambridge University Press

Fisher, S. C., J. A. Hurst, F. Varga-Khadem, and A. P. Monaco. 2001. A forhhead-domain gene is mutated in a severe speech and language disorder. Nature 413, 519–523.

Gilbert, D. 2004. Bioinformatics software resources. Breif. Bioinform. 5, 300–304.

Kidwell, E. 2005. Intelligent Bioinformatics: The Application of Artificial Intelligence Techniques to Bioinformatics Problems. New York: Wiley.

Kohane, I. S. 2002. Microarrays for Integrative Genomics. Cambridge, MA: MIT Press.

Nair, A. S. 2007. Computational Biology & Bioinformatics—A Gentle Overview. Secundirabad, India: Communications of Computer Society of India.

Olson, S. 2007. Mapping human history: Discovery the past through our genes. Boston, MA: Houghton Mifftin.

Pollard, K. S. et al. 2006. An RNA gene expressed during cortical development evolved rapidly in humans. Nature 443, 167–172.

Saltzberg, S., D. Searls, and S. Kasif. 1998. Computational Methods in Molecular Biology. Amsterdam, The Netherlands: Elsevier.

CHAPTER 3

METHODOLOGY FOR SEPARATION AND IDENTIFICATION OF PROTEINS AND THEIR INTERACTIONS

In this chapter, we will discuss methods for (a) the isolation and separation of proteins, (b) the identification of the primary structure of proteins via determination of their amino acid sequences, (c) the 3D structure, of proteins and (d) the amount of proteins at the proteomic scale.

3.1 SEPARATION OF PROTEINS VIA A MULTIDIMENSIONAL APPROACH

A cell contains a myriad of proteins. The individual proteins are separated and identified to understand the nature of different proteins and their structural and functional relationships. Proteins are isolated by a variety of means because they have a diverse set of properties. A multidimensional approach is more suitable to separate them than a single approach, as outlined in Chapter 1. Such a multidimensional approach must address the problems concerning their resolution, high throughput, automation, and adaptability to analysis by mass spectrometry. One of the most important ways they are separated is by electrophoresis, including the two-dimensional (2D) gel electrophoresis and capillary electrophoresis. After their separation by these methods, they are identified by use of a mass spectrometry. These methods are described in this chapter.

Introduction to Proteomics: Principles and Applications, By Nawin C. Mishra

3.1.1 Electrophoresis

Molecules can be separated in an electrical field based on their electrical charges. This method is called "electrophoresis" and was developed by Swedish scientist Arne Teselius; for this development, he was later awarded the Nobel Prize in 1948. Molecules are applied on a solid support such as paper or gel and then subjected to an electrical field. The molecules move in an electrical field depending on their electrical charges. Proteins are usually separated on a gel matrix as discussed below.

3.1.1.1 2D Gel Electrophoresis. This method of separation of proteins was developed independently by O'Farrell (1975) and by Klose (1975) (Figure 3.1). The Proteins present in the cell extract of *Escherichia coli* were separated by gel electrophoresis run in two different planes at right angles to each other. First, the proteins were separated based on their charges, and then they were separated on the basis of their molecular weights in a plane at a right angle to the plane of the first separation. The gel was stained after the succession of two electrophoretic runs to visualize the protein bands. In this way, more than 1100 protein bands were obtained from the *E. coli* total cell extract. The 2D gel method also has been called the Iso-Dalt method, because in this method proteins are separated based on their differences in electrical charge and mass as expressed, respectively, by isoelectrofocusing (IEF) [isoelectric point (pI)] and Dalton, the unit of mass.

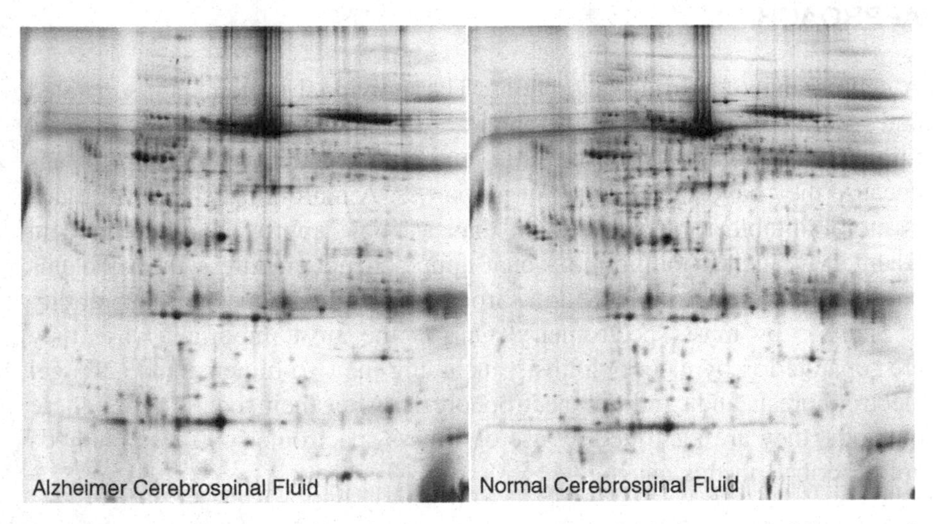

Figure 3.1: 2Dgel separation of proteins in cerebrospinal fluids from normal human and Alzheimer patient. (Picture reproduced with the permission of Professor Kelvin Lee, University of Delaware.)

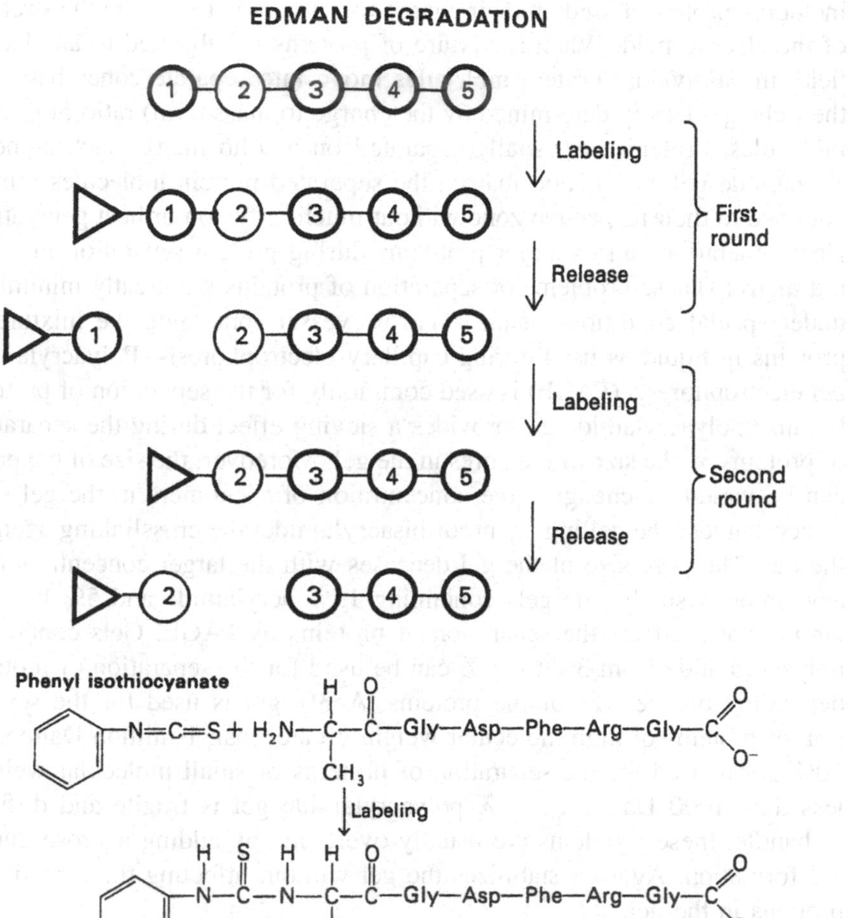

EDMAN DEGRADATION

Labeling

First round

Release

Labeling

Second round

Release

Phenyl isothiocyanate

Labeling

Release

Peptide shortened by one residue

PTH-alanine

Figure 3.2: Steps in Edman degradation. The labeled aminoterminal residue (PTH - alanine in the first round) is released without hydrolyzing the remainder of the peptide. The cycle is repeated to reveal the complete sequence of the peptide. (Reproduced from *Biochemistry* by L. Stryer, K.B. Freeman, and company with the permission of the publisher.)

Electrophoresis is a common method for separating different molecules, including proteins based on their charge-to-mass ratio (e/m) and the strength of the electric field. When a mixture of proteins is subjected to an electric field, the individual protein molecules move into separate zones based on their charge density determined by the charge to mass (e/m) ratio of protein molecules. Proteins are usually separated on a solid matrix such as polyacrylamide gel. On a solid matrix, the separated protein molecules remain confined to their respective zone without much diffusion or heat generation. Heat generation causes major problems during protein separation in a liquid matrix. These problems of separation of proteins are greatly minimized under special conditions using a narrow vessel containing the mixture of proteins in liquid as used during capillary electrophoresis. Polyacrylamide gel electrophoresis (PAGE) is used commonly for the separation of proteins because polyacrylamide gel provides a sieving effect during the separation of proteins by the size of the pores in the gel. Moreover, the size of the pores can be varied by changing the concentraion of monomers in the gel such as acrylamide, the gelling agent, or bisacrylamide, the crosslinking agent in the gel. The pore size of the gel deceases with the larger concentration of acylamide. Usually, the gels containing 15% acrylamide and 5% bisacrylamide are used for the separation of proteins by PAGE. Gels containing polyacrylamide from 3% to 30% can be used for the separation of proteins depending on the size of the proteins. A 3% gel is used for the separation of proteins of high molecular weight greater than 1 million Daltons. A 30% gel is used for the separation of proteins of small molecular weights less than 1000 Daltons. A 3% polyacrylamide gel is fragile and difficult to handle; these problems are usually overcome by adding agarose during gel formation. Agarose stabilizes the gel without affecting the mobility of proteins in the gel.

The 2D gel or Iso-Dalt is the most commonly used method in proteomics because of its relatively easy use, automation, high reproducibility, high resolution of proteins, and applicability to analysis by mass spectroscopy. Furthermore, protein bands separated by Iso-Dalt are readily amenable to Edman degradation or to the amino acid composition analysis.

As mentioned, the method of 2D gel uses the power of separation of proteins by resolution based on charge and mass of the proteins in two separate dimensions. A separation based on this dual property of proteins provides high resolution by avoiding the presence of several proteins per band and or cross-contamination of a protein band by other protein components and by visualizing the proteins that occur in a small amount in the cell or body fluid. To attain high resolution of thousands of proteins in proteomics, it is important to start with a protein sample that includes all proteins present in a cell or body fluid. Attention must be paid to include all membrane proteins or other hydrophobic proteins and rare protein components that occur in small copy number as well as proteins with extreme differences in their charge content, such as those below pH 3 or above pH 10, present in the cell or body fluid.

The 2D gel method for the separation of proteins has been described in detail by Gorg and others (1988), and Gorg (2000). The important aspects of this methodology consist of the following steps:

1. Sample preparation, solubilization, and application on to gel
2. Separation of protein by IEF on an immobilized pH gradient (IPG) gel strip
3. Separation of proteins present in the IPG strip on a sodium dodecyl sulfate (SDS)-PAGE
4. Visualization of separated protein bands after SDS-PAGE and evaluation of the pattern of the separated proteins on gel after visualization

3.1.1.1.1 Sample Preparation, Solubilization, and Application on to Gel. Cells, tissue, organs, or their TCA acetone extracts are kept deep frozen until ready to prepare them for use in 2D gel. Usually, the frozen samples are first disintegrated by a variety of means, including crushing or grinding in a mortar with a pestle in the presence of liquid nitrogen, and then homogenized by shearing in a waring blender or in a sonicator. Such extracts are then solubilized by sonication in the lysis buffer containing urea or thiourea, dithiothreitol (DTT), a detergent, a cocktail of protease inhibitors, and carrier ampholytes. The samples that are prepared as TCA-acetone extracts are solubilized directly in lysis buffer without homogenization. In addition, specific methods are developed for the extraction of certain protein samples, such as plant leaf proteins or ribosomal proteins and histones. In the solubilization buffer, urea and thiourea work as chaotrope and help in solubilization by disruption of hydrogen and hydrophobic bonds without any change in the intrinsic charge of the protein. The presence of urea may cause some carbamylation of proteins, which is controlled by the adjustment of pH, temperature, and ionic strength of the solubilization buffer. Carbamylation involves the addition of cyanate to protein molecules. DTT in the solubilization buffer acts as a reductant by disrupting disulphide bonds and helps in the denaturation of proteins. Detergents help in the disruption of membranes and removal of lipids from proteins bound to vesicles. The inclusion of protease inhibitors in the solubilization buffer prevents the degradation of proteins by the proteases intrinsically present in the sample. Ampholytes help in many ways by maintaining the pH and removing cyanate ions responsible for the carbamylation of proteins. It reduces the interaction between proteins and immobilins in the gel strip.

However, the ampholytes interact with silver and Coomassie stains used for the visualization of proteins in gel and must be removed after electrophoresis before the visualization of protein bands.

After solubilization of proteins in the sample lipids, nucleic acid and polysacharides are removed by ultacentrifugation. Lipids usually float at the top and are removed. Nucleic acids and polysacharides sink to the bottom of the tube and are discarded by gently pouring off the liquid containing the proteins into another tube. The removal of lipid, nucleic acid, and carbohydrate is important, as these elements clog the gel pore and also interact with proteins and cause changes in migration of proteins through the gel. The presence of high salt levels is a problem in the uniform flow of electric current during the electrophoresis and should be avoided before the preparation of sample. One way to avoid high salt is performing dialysis on the sample before lyophilization.

3.1.1.1.2 Separation of Proteins in the First Dimension Based on Charge Alone by IEF. Proteins after solubilization and removal of all interfering materials are applied to the top of a premade polyacrylamide gel strip containing ampholytes or ampholines with a pH range of pH 6–10. During electrophoresis, the mixture of ampholines becomes distributed to specific zones based on their charges and creates zones with a pH gradient in the gel strip varying in pH from 6 to 10. Proteins migrate to a particular zone of the gel based on their isoelectric point where their net specific charge reaches zero. Thus, proteins are separated from one another based on their charge alone, irrespective of their mass. Ampholines or ampholytes are amphoteric substances carrying both positive and negative groups. After electrophoresis, ampholytes form zones of a specific pH gradient, but they are subject to electroosmotic flow, which causes their movement to the cathodic end of the gel. Such diffusion of ampholytes may cause irreproducibility of the protein separation from one experiment to another. The pH gradient is, however, stabilized by having certain substances called immobilines that have week buffering capacity; they are attached to the matrix of the gel so they cannot move around and yield gels of IGP. Usually, a sample containing up to 4 mg of proteins is applied to the top of a gel strip of 18 cm in length and subjected to electrophoresis at an optimal temperature of 20°C.

3.1.1.1.3 Separation of Proteins in Second Dimension Based on their Mass Alone. The gel strip containing the proteins separated by electrofocusing is equilibrated in the SDS buffer and then placed on a slab of polyacrylamide gel with SDS and electrophoresed to separate the proteins. All proteins after interaction with SDS become negatively charged, and thus, they are separated based on their mass alone.

3.1.1.1.4 Visualization of Proteins in Gel After Electrophoresis. Proteins are visualized as bands or spots on the gel by staining with a dye. The different dyes for the staining used include (a) Zinc or copper, (b) Coomassie Blue, (c) Silver, and (d) Fluorescent dye. Zinc or copper staining is a negative staining because it stains the gel and not the protein spots that are covered with SDS. This method is inexpensive and has a sensitivity to detect spots containing 6–12 ng of protein, but it has difficulty in handling thin gels. Coomassie Blue is an easy and an inexpensive stain. It is used to identify proteins by mass spectrometry. Is has a sensitivity of 36–45 ng of protein per spot in the gel. Silver staining is expensive and time consuming but worth its sensitivity. It can detect 0.5–1.2 ng of protein per spot in the gel.

Fluorescent dye comes in a variety of choices; they are quick and easy to use and are highly sensitive like Silver stain but not compatible to subsequent techniques of protein identification by mass spectroscopy.

The methodology for 2D gel is now completely automated. Robots are used to pick up a particular protein spot from the gel and are made available for analysis by mass spectrometry. All equipment and different solutions used for 2D gel analysis of proteins are commercially available. Despite all the advantages of the 2D gel system in proteomics, this methodology still suffers from several drawbacks. Some of these drawbacks include the lack of the detection of low abundance proteins such as receptor, regulatory, and signal transduction proteins. Also, basic proteins as well as the membrane proteins that represent 40% of all cellular proteins are hard to separate by 2D gel. Some of these drawbacks can be overcome by running several gels with 1 pH difference, at pH range higher than 10, or by prefractionation of proteins before 2D gel.

3.1.1.1.5 2D Differential In-Gel Electrophoresis (DIGE). DIGE this is a modification of 2D gel electrophoresis to avoid any differences that are usually encountered when samples are run on different gels even under identical conditions. In this method, two protein samples with the same amount of protein in each are linked covalently at lysine residues with fluorescent cyanine dyes cy3 and cy5. The two samples are then mixed and run on the same gel, and then the separated proteins are visualized by excitation of cy3 or cy5 specific fluorescence imaging. The differences in the protein bands of normal and cancer cells have been visualized in this way.

3.1.1.1.6 Recent Advances in the Imaging of 2D Gel Electrophoresis. Progress has been made for better imaging of protein bands on the gel by the development of a process called “diversity.” In this method, the image of fluorescent protein bands in the gel is captured in a single shot with a sensitive 6.2-pixel camera with high resolution. This process alleviates the need for stitching the images of the gel band by combining several pictures of the protein band in a gel. This process offers a wide variety of filters for the imaging of protein bands stained with dyes of different colors. In addition, a new software package called "Dimension" has been developed for rapid analysis of the protein band after electrophoresis.

3.1.1.2 Capillary Electrophoresis. Capillary electrophoresis involves the separation of proteins and peptides carried out in a thin glass tube usually 50 μm in diameter under high voltage. Thin tubes have the advantage of dissipating heat by high voltage, and separated proteins can be visualized and monitored by ultraviolet (UV) light during the eletrophoretic run. The capillary method can be run using a gel or isoelectrophoretic system. Capillary electrophoresis is generally used to separate peptides before injection into the mass spectrometer for their identification.

3.1.2 Liquid Chromatography

Chromatography is the process of separation of proteins or peptides by passing a solution of proteins/peptides in an appropriate solvent over a solid matrix. Depending on the nature of the matrix used, there are different kinds of chromatography such as liquid or column chromatography, paperchromatography, thin gel chromatography, and gas chromatography. Among these different kinds of chromatography, the one that is most relevant for proteomic studies and suitable for analysis by mass spectroscopy is liquid chromatography; this is also known as column chromatography, because the matrix consists of beads packed as a column in a glass tube over which the protein or peptide solution is passed, partitioned based on their size, charge, or affinity to a ligand, and collected as fractions of separated proteins or peptides.

3.1.2.1 Gel Filtration. Liquid chromatography that separates proteins based on their size is called size-exclusion chromatography or gel filtration. In this method, proteins are separated based on their molecular size. A solution of proteins is passed over a column packed with inert substances of a fixed pore size such as agarose or sephedex, and then eluted with a buffer. During this process, the larger proteins that cannot enter into the pore of the beads snake through the space between the beads and are excluded from the column during the process of elution. Thus, larger proteins come out first, much before the smaller proteins. In contrast, the smaller proteins enter the bead pore and take a much longer time to come out of the pores during elution, and these proteins are eluted much later than the larger proteins. In this way, the different-sized proteins are separated and collected as distinct fractions after elution.

3.1.2.2 Affinity Chromatography. Liquid chromatography that separates proteins based on their affinity for a ligand attached to the matrix is called affinity chromatography. In this process, proteins that lack any affinity for the ligand attached to the matrix remain unbound and are removed readily from the column during the process of elution. In contrast, a particular protein in the mixture may bind to a ligand and is delayed in their removal from the column; this protein is removed later from the column with change in the conditions of elution. This chromatographic method is used to separate proteins from the bulk of a protein that remain bound to the ligand attached to the column. A good example of purification of protein by affinity chromatography includes certain proteins containing histidine oligopeptides that are removed on a nickel column. Also, proteins containing glutathione-S-transferase can be retained and later removed selectively from a column containing beads coated with glutathione. An affinity tag can be custom designed to isolate or purify a particular protein. For example, proteins that are involved in DNA transactions, such as DNA replication, repair, and recombination, can be purified over a matrix containing DNA attached to it as affinity tag. Alternatively, a protein can be purified by passing over a matrix containing an antibody to this protein attached as an affinity tag.

3.1.2.3 Ion Exchange Chromatography. Liquid chromatography that separates proteins based on their charges is called ion exchange chromatography. In this process, proteins are adsorbed into a charged group attached to the cellulose matrix. The charged groups are the organic molecules that are either anionic or cationic. Carboxymethyl (CM) cellulose and diethylaminoethyl (DEAE) cellulose are the most commonly used cation and anion exchanger used as a matrix during ion exchange chromatography.

The unbound proteins come out first, and later the bound proteins are removed from the column with a change in the ionic concentration or pH of the elution buffer.

3.1.2.4 Reverse-Phase (RP) and High-Performance Liquid Chromatography (HPLC). Reverse-phase chromatography separates proteins based on the hydrophobicity of the molecules. It results in the separation of molecules based on the mass of proteins, because hydrophobicity of the molecules is determined by the mass of the molecules. In this chromatography, the matrix consists of ligands that bind reversibly with the hydrophobic proteins. The matrix-bound proteins are then removed by appropriate change in the elution conditions; proteins of lower hydrophobicity are removed first. Most of the time, the protein mixture and their subsequent elution is forced through the matrix under high pressure; this leads to the high performance of the chromatographic process. Therefore, this method is called HPLC. HPLC is well suited to the delivery of protein sample to the mass spectrometer, which is maintained under high vacuum.

3.1.2.5 Multidimensional Chromatography. In proteomics, the separation of proteins is usually accomplished by applying a combination of chromatographic methods. Most proteins are first fractionated by ion exchange chromatography and are separated based on their charges. Subsequent to this, they are subjected to separation by RP-HPLC based on their mass. This combination of ion exchange chromatography and HPLC has the same effect as the 2D gel electrophoresis on the separation of proteins. This combined approach is also called multidimensional chromatography, as it separates proteins in two different dimensions based on their charges and mass in quick succession, which is done in the 2D gel electrophoresis. This approach has the advantage of separating certain membrane proteins that are not amenable to 2D gel analysis. However, this lacks in that it cannot provide certain information like the pI of the proteins during their analyses by comparing data in the gene bank and protein bank.

To enhance the resolution of multidimensional chromatography, the effluent from the ion exchange chromatography is passed through a series of HPLC columns and then injected into mass spectrometer.

At times, the effect of separation by charge and by mass is combined in the same column with the use of a biphasic matrix; the distal half of the column is filled with HPLC resin for separation based on the mass of proteins, and the proximal half of the column is packed with ion exchange matrix separating the proteins based on the charge. This biphasic column is used when the elution buffer for the two systems is the same or compatible. Such separation by a biphasic column is suitable for the multidimensional protein identification technology (mudPIT) by mass spectroscopy.

3.2 DETERMINATION OF THE PRIMARY STRUCTURE OF PROTEINS

It is the sequence of amino acids that constitutes the primary structure of a protein. This in turn determines the way a protein folds and assumes the three-dimensional (3D) structure that controls the activity of a protein. Many other inferences could be made from the amino acid sequence of a protein; for example, its (pI) can be deduced from the amino acid sequence of a protein. Also, the presence of many hydrophobic amino acids in the sequence would indicate that it is a membrane protein or that it may be a receptor protein. Also, the presence of certain amino acids may indicate that it would form a betasheet in the protein structure. Thus, the sequence of amino acids in a protein and its primary structure is important. There are three ways to determine the amino acid sequence of a protein. These include deciphering from the nucleotide sequence of a DNA molecule, Edman degradation, and mass spectrometry.

3.2.1 Proteomics without Spectrometry

3.2.1.1 Determination of Amino acid Sequence from DNA sequence. Understanding genetic codes and our ability to sequence nucleotides in a DNA segment made it possible to decipher the amino acid sequence of a protein encoded by a gene. It became customary to infer the amino acid sequences of different proteins as soon as the DNA sequence data became available in the gene bank. The primary structure of a large number of proteins deposited in the protein data bank (PDB) is obtained directly from the DNA sequence data.

Although it is easy to infer the amino acid sequence of a protein from the nucleotide sequence, the converse is not true (i.e., it is not easy to infer the nucleotide sequence of a gene from the amino acid sequence of a protein because of the degeneracy of the genetic code). To infer the DNA sequence of a gene based on the amino acid sequence of a protein, one has to rely on the common usage of genetic code for that organism to which the gene belongs. It is common practice to construct a gene for cloning in molecular biological experiments. The first insulin gene was constructed based on the amino acid sequence of insulin protein in this manner. In prokaryotes and lower eukaryotes, such as yeast and many filamentous fungi, the amino acid sequence could be deciphered directly from the nucleotide sequence. However, in higher eukaryotes, such as mammals, there is no continuous correlation between the nucleotide sequence and the amino acid sequence because of the presence of intron sequences. Thus, the intron sequences must be ignored while deciphering the amino acid sequence of a protein encoded by a gene in higher organisms. Furthermore, in higher organisms, many proteins can be theoretically formed depending on the manner of splicing of the exons. In nature, not all these theoretically possible proteins may occur in an organism.

At times, the availability of the nucleotide sequence of gene encoding a protein is crucial in deciding the amino acid sequence; For example, the occurrence of two adjacent glycine residues, each with a molecular mass of 57 daltons, in a peptide may appear as asparagines having a molecular mass of 114 daltons in mass spectrometric analysis. However, this situation can be resolved easily by examining the nucleotide sequence showing different genetic codes for glycine and asparagine. Such examination of the nucleotide sequence can establish conclusively whether there are indeed two adjacent glycine residues or just an asparagine residue in the amino acid sequence of a protein in question.

3.2.1.2 Edman Degradation — N-Terminal Amino Acid Sequence Analysis. Before mass spectrometry became available, the sequence of amino acid was determined by Edman degradation. This involved the identification of one amino acid at a time from the N-terminus of the peptide. The first N-terminal amino acid is reacted with phenyl-iso-thio-cyanate, and then the N-terminal amino acid is cleaved away by mild hydrolysis as a cyclic complex of the N-terminal amino acid called phenyl hydrothion. This process leaves the shortened peptide intact. The cleaved amino acid is identified by its chromatographic profile. Then, the second N-terminal amino acid is cleaved, and the remaining peptide shortened by two amino acids is again obtained intact. The cleaved amino acid is identified by its chromatographic profile. This process is repeated after every cycle until all the amino acids are identified. This process is automated completely. However, it is laborious and time consuming. Despite these inherent difficulties, this method, which is called Edman degradation and was developed by Pehr Edman, was the only method before the application of mass spectrometry to determine the amino acid sequence of a protein. The steps in Edman degradation are presented below.

When the methodology of 2D gel for separation of proteins became available in the mid-1970s, the proteins separated by 2D gel were routinely subjected to Edman degradation to determine the amino acid sequence, because this was the only way to determine amino acid sequence of a protein at that time. However, there are two problems with this approach in proteomics: First, many proteins possess N-terminal proteins that are blocked and therefore cannot react with the phenylisothiocynate required for Edman degradation. Second, Edman degradation can determine the amino acid sequence of one protein at a time, which is contrary to the objectives of proteomics that aim at gaining the information of several proteins at the same time.

3.2.2 Proteomics Based on Mass Spectrometry – Identification of Proteins Based on Their Amino Acid Sequence

Proteins or peptides after separation by various methods, such as electrophoresis and/or liquid chromatography, are identified by determining the sequence of amino acids comprising them. Traditionally, this was done by Edman degradation, which determined one amino acid at a time from the N-terminus of the proteins or peptides. However, the process of the identification of proteins was revolutionized by the development and application of the mass spectrometer in conjunction with the advances in genomics and bioinformatics, which made the gene and protein data available for the assignment of a particular peptide sequence to a protein and to the encoding gene.

3.2.2.1 Mass Spectrometry.

3.2.2.1.1 Principle, Historical Perspectives, and Instrumentation. The spectrometer works on the principle that molecules can be ionized, and the ionized molecules can be separated based on their mass-to-charge ratio. The results yield information about their molecular weights and chemical structure. The principles of the mass spectrometer originated in the middle of the 19th century from the study of gas discharge in an electric field, which yields cathode and anode rays. Almost 50 years after the discovery of the cathode and anode rays, Joseph J. (J.J.) Thomson in the Cavendish laboratory of the Cambridge University in England observed that in an electrical field, molecules can give rise to positively and negatively charged particles corresponding to cathode and anode rays. The positively charged particles are the ionized molecules, whereas the negatively charged particles are the electrons. It was also found that these charged particles could be separated based on their mass-to-charge ratios by applying a magnetic force. These findings led to the development of the first mass spectrometer by Francis Aston, who was a student of J. J. Thomson. Based on these properties of charged molecules, first Thomson and later Ashton were able to establish the existence of the different isotopes of a stable element and the internal organization of the different components, such as protons, neutrons, and electrons in the atom of an element. This in turn led to a better understanding of the physics and chemistry of the chemicals in nature and ushered the coming of particle physics as a new branch of science. This also led to the discovery and creation of new elements. These also led to the discovery of the components of the subatomic particles. In contrast to these developments in the academic aspects of the science regarding the nature of particles, the findings of J. J. Thomson led to more practical applications such as the development of nuclear power both in terms of the making of nuclear weapons and the generation of nuclear energy. The use of mass spectrometers was crucial for the separation of uranium isotopes for the making of the first atomic bomb during Manhattan Project under the leadership of Robert Oppeheimer in 1945. For a period of 100 years beginning with the original findings of J. J. Thomson, several scientists were awarded Nobel Prizes in physics or in chemistry on research leading to the development of a spectrometer. Some of these included J. J. Thomson, Francis Ashton, Wolfgang Paul, John Fenn, and Koichi Tanaka. Mass spectrometers have been used for the separation and identification of a variety of chemicals. However, it was the development of ionization processes developed by John Fenn and Koichi Tanaka that made the analyses of proteins and peptides feasible for its use in proteomics.

3.2.2.1.2 Instrumentation. The mass spectrometer is a sophisticated instrument that measures the molecular weight of a substance with a great precision and helps to identify the chemical nature of that substance. This machine ionizes molecules and separates them according to their mass/charge (m/z) ratio and yields information regarding the molecular weight of each ion. This information is used to determine their structure. Each amino acid has a distinct molecular weight; this in turn gives each peptide a characteristic molecular weight. This characteristic molecular weight of a peptide is used to decipher the sequence of amino acids in a peptide or in a protein. Even a subtle difference in the molecular weight is an indication of replacement of an amino acid by another amino acid or of a posttranslational modification, including phosphorylation, acetylation, or other changes in the structure of a peptide. The molecular weights of different amino acids are presented in Table 3.1.

It should be noted that the molecular weight of amino acids in a peptide is less than the molecular weights of the free amino acids. It is decreased by 18 daltons, which is the molecular weight of water, because a molecule of water is removed during the peptide formation. The mass spectrometer is of great use in proteomics. This instrument is used to decipher the nature of a protein primarily its amino acid sequence and protein ligand complex formation under physiological conditions, including posttranscriptional modifications, enzyme-substrate bindings, and antigen–antibody or orphan–receptor interactions. This instrument also is used to understand the nature of protein folding by monitoring the hydrogen/deuterium exchange.

3.2.2.1.3 Components of the Instrument. A spectrometer consists of the following five major components: a port or device for the introduction of

Table 3.1. Molecular weights of amino acids in peptides.

Amino acids	Symbols—3 letter/1 letter	Molecular weight
Alanine	ala/A	71
Asparagine	asn/N	114
Aspartate	asp/D	115
Arginine	arg/R	156
Cysteine	cys/C	103
Glutamine	gin/Q	128
Glutamate	glu/E	129
Glycine	gly/G	57
Histidine	his/H	137
Isoleucine	ile/I	113
Leucine	leu/L	113
Lysine	lys/K	128
Methionine	met/M	131
Phenylalanine	phe/F	147
Proline	pro/P	97
Serine	ser/S	87
Threonine	thr/T	101
Tryptophan	trp/W	186
Tyrosine	tyr/Y	163
Valine	val/V	99

sample into the machine, a device for ionization of molecules, an analyzer for the separation of ionized molecules on the basis of their mass to charge (m/z) ratio, a detector that monitors the presence of the separated ions and records them, and a high vacuum system to allow free movement of ions within the spectrometer.

Device for the introduction of sample into the machine. A sample may be injected into the spectrometer depending on the nature of the sample and the method of ionization. A sample is placed into the source of ionization in the spectrometer. It is usually introduced on a probe or a platform or via a capillary tube directly after the HPLC or capillary electrophoresis of proteins or peptides. A sample is always introduced through a lock system to maintain the high vacuum in the machine without any interruption.

Ionization device. In a spectrometer, molecules are subjected to ionization because it is easier to control and direct the movement of ionized molecules with an electrical charge than neutral molecules. Usually, the ionization is achieved either by protonation (addition of an H+ ion) or by deprotonation (removal of an H+ ion). These two methods are called positive and negative ionization. Proteins and peptides are usually subjected to positive ionization because the NH_2 group in protein readily accepts an

H+ ion. Several methods of ionization of molecules are introduced into the machine for analysis. However, two major kinds of devices for the ionization are used in proteomics. These include the eletrospray ionization (ESI) device and the matrix assisted laser desorption ionization (MALDI) device. These devices are described in the following sections.

3.2.2.1.4 Electrospray Ionization. Electrospray ionization was developed by John Fenn (1989). Fenn was awarded the Nobel Prize for this work. Electrospray ionization is a soft ionization process because it does not disintegrate the molecules. In this method, a sample is vaporized by high voltage and then ions are generated as the solution of proteins or peptides is forced through a fine syringe. This vaporization process is similar to a spray of perfume coming out of a bottle when the tip is pressed. In case of the perfume, the jet of spray is created by the air that rushes in when the tip of the bottle is pressed. In the case of electrospray ionization, a jet of vaporized molecules is created by high-voltage electricity, and the molecules become smaller and smaller as they move away from the tip of the syringe to form a conical-shape aerosol spray called the Taylor cone.

During electrospray ionization, analyte (i.e., the material such as proteins or peptides to be analyzed by the mass spectrometry) is dissolved in a volatile organic solvent and then forced though a fine metal capillary needle charged with a high voltage of electricity. Usually, solvent containing the analyte in the amount of 1 μL/minute to 1 ml/minute is forced through the charged capillary needle. The solvent containing the ionized analyte is vaporized and comes out as a Taylor cone. In this cone, as the solvent dries up, the bubbles become smaller, and finally the bubbles just contain the ionized analyte molecules. The process for drying of solvent is sped up by injecting heated neutral gas such as nitrogen. The ionized molecules in the bubble contain the same charges that therefore repel each other and finally overcome the surface tension. The bubbles containing them burst open, and the ionized molecules become free. These ionized molecules are moved toward the detector by the analyzer based on their mass-to-charge ratio. As a result of electrospray ionization, the ionized molecules may contain one or more charge(s). The protonated ionized molecules are designated as M + H, M + 2H, or M + Hn depending on one, two, or more positive charges. Likewise, the deprotonated molecules containing one, two, or more negative charges are designated as M-H, M-2H, or M-Hn, respectively. A sample of proteins or peptides in a suitable buffer can be readily dissolved in an organic solvent for the process of electrospray ionization. This fact makes the effluents coming out of LC/HPLC more suitable for analysis by mass spectrometry. This adaptability of LC/HPLC has revolutionized the seperation and identification of proteins and peptides and has advanced the science of proteomics. A recent version of electrospray ionization called "nanospray ionization" has become more popular. In nanospray ionization, a much smaller volume of liquid as little as 1 nL/minute is passed through the charged capillary needle. This results in generation of a finer spray with much reduced size of the ionized droplets.

MALDI is another equally important method of ionization that is used in proteomics. This process was first developed in 1988 by Koichi Tanaka in Japan and was later improved by Karas and Hillenkamp (1988) in Germany. Tanaka was awarded the Nobel Prize for this work. In this method, an analyte is adsorbed onto the surface of an UV-absorbing matrix. The adsorption of analyte by the matrix is achieved by cocrystalization of analyte and matrix in the molar excess of the matrix material. The matrix containing the analyte is then exposed to the pulse of a nitrogen laser beam, which vaporizes the analyte ionizing the analyte molecules. Thus, during this process, the matrix serves both in absorbing the energy and in donating proton leading to the vaporization and ionization of the analyte. The process results in soft ionization and does not cause fragmentation of the analyte molecules. MALDI usually leads to singly charged molecules, which are easily managed during the subsequent sets of analysis by a mass spectrometer. Analytes containing proteins and peptides are positively charged by MALDI. MALDI causes negative ionization of oligonucleotides and carbohydrate molecules. The positive ionization is usually promoted by the inclusion of a trace amount of formic acid, whereas the presence of ammonia during the ionization process causes negative ionization. The process of ionization by MALDI is crucial to the development and application of the mass spectrometer in proteomics. MALDI is the cornerstone of the identification of whole proteins or peptides by providing accurate information regarding their mass in a spectrometer-based proteomic analysis.

3.2.2.1.5 Other Methods of Ionization. There are several other methods for ionization in addition to ESI and MALDI. However, most of them are not commonly used in proteomics. Some of these include chemical ionization, electron ionization, fast atom bombardment (FAB), and many others. Most of these lead to disintegration or fragmentation of analyte molecules and are not commonly used in proteomics. However, FAB has some application in the analysis of proteins and peptides, because this is a soft ionization procedure and does not cause the fragmentation of molecules under analysis. In the FAB method, a nonvolatile matrix such as m-nitrobenzyle alcohol is used to hold the analyte molecules. Analyte molecules are vaporized and ionized by bombardment with the high-energy beam of xenon or cesium from a probe inserted directly into the device containing the sample. Ionized molecules thus obtained are then subjected to separation by the mass analyzer based on mass-to-charge ratio as in other methods of ionization, including ESI and MALDI.

The mass analyzer. After the process of ionization, the ionized molecules of proteins or peptides enter the section of the mass spectrometer called the mass analyzer, where they are separated based on their mass-to-charge ratio by electric and/or magnetic fields or by measuring the time taken by an ion to reach a fixed distance from the point of ionization to the detector. Different kinds of analyzers are available for the separation of ionized molecules. Among the different kinds of analyzers, two particular kinds, called the quadrupole and the time-of-flight (TOF) analyzers, are the most important from the point of proteomics for their use in mass spectrometers. A particular spectrometer may use one or the other or at times a combination of both quadupole and TOF analyzers. Usually, the machine with the electrospray ionization device carries a quadrople analyzer. A spectrometer with a MALDI device has a TOF analyzer or a combination of quadrupole and TOF analyzers in succession to each other. Certain spectrometers called "tandem spectrometers" (MS/MS) contain two or three quadrupoles and a TOF analyzer.

3.2.2.1.6 Quadrupole Mass Analyzer. This type of analyzer was devised by Wolfgang Paul. It contains four rods with a direct current (dc) voltage that is superimposed with a particular radio frequency (RF) to select ions within a particular mass range. This device allows ions of a particular mass range to reach the detector while deflecting other ions of different mass range away from the detector. The quadrupole mass analyzer is an inexpensive analyzer and is used in a spectrometer with an electrospray ionization device. It has the advantage to work at atmospheric pressure at which the electrospray ionization is usually carried out. Thus, it does not have a rigorous requirement for an ultra-low vacuum system in the machine. This mass analyzer is good for the analysis of peptides up to molecular weights of 3000 daltons, but it is not suitable for the analysis of proteins, which are usually of molecular weights much greater than 3000 daltons.

3.2.2.1.7 TOF Mass Analyzer. This analyzer is used in conjunction with the MALDI device. This analyzer works on a very simple principle that involves accelerating a number of ions at the same energy level so that the ions of different masses take different time lengths to reach the detector. The ions with smaller mass reach the detector much faster than the ions with a larger mass. This situation can be likened to the case of a pitcher hitting a golf ball or a baseball. The golf ball being lighter will reach the catcher much faster than the baseball, which is comparatively much heavier than the golf ball.

3.2.2.1.8 Other Mass Analyzers. Other analyzers, such as quadrupole ion trap (QIP) and Fourier transform mass spectrometer (FTMS), are of some interest for proteomics. The quadrupole ion trap mass analyzer was devised by Wolfgang Paul; it works on the principle of trapping ions with a particular RF in the quadrupole mass analyzer. This device provides a way to eject ions of certain radio frequency and retain the others, only the latter are allowed to reach the detector by scanning ions of a particular radio frequency. In this method, the selected ions can be subjected to fragmentation by collision-induced dissociation (CID), which is useful for the analysis of peptides.

Fourier transform mass spectrometry was devised by Comisarow and Marshall in 1974 at the University of British Columbia. This method uses Fourier transformation, which is a mathematical way to convert a set of data involving a time domain into a set of frequencies. The mathematical concept of Fourier transformation is important and has been used in analyses of the protein structure by mass spectrometry, X-ray crystallography, and nuclear magnetic resonance (NMR).

The method of Fourier transform mass spectrometry works on a simple principle of exciting ions, which orbit in a magnetic field, by a radio frequency signal to produce detectable image current. The latter is then Fourier transformed to obtain the frequency of the component ions based on their mass-to-charge ratio. This method has great resolution with an accuracy of 0.001%.

Detector. This is the final component of a mass spectrometer. Its purpose is to detect and record the presence of ions coming out of the mass analyzer hitting the detector. An electron is emitted when an ion hits the recorder and creates a small current. The low level of signal from a small number of ions coming out of the mass analyzer is amplified from 1000 to 1 million times to become delectable and then recorded. A detector may use an electron multiplier or a photomultiplier. Photomultipliers first convert an electron produced by the ion hitting the detector plate into photon, which is detected by a phosphorescent plate in a sealed tube. Photomultipliers are preferred in a detector because they are located in a sealed tube, which reduces the noise-to-signal ratio by not allowing any outside interference to come out from the mass analyzer. All mass spectrometer are equipped with photomultipliers. These signals are then recorded on a graph by plotting the amount of signal versus m/z ratio. Mass spectrometer graphs usually show the presence of proteins/peptides of different molecular size and their abundance. The graph may also show the energy level of the molecules of a particular kind.

Vacuum system. A mass spectrometer is always operated under a low vacuum. Electrospray ionization, however, has a low vacuum requirement and may be carried out at or near atmospheric pressure.

3.2.2.1.9 Tandem Mass Spectrometer (MS/MS). A tandem mass spectroscopy involves two or more mass spectrometers in succession connected to one another. Alternatively, a single mass spectrometer contains several analyzers arranged one after another. Both of these arrangements provide multiple methods for the selection and/or analysis of proteins/peptides. In addition, the tandem arrangement may have a provision for the fragmentation of proteins or peptides if so required. The fragmentation of a selected protein may be achieved by various ways such as CID or electron capture dissociation (ECD). Thus, tandem mass spectrometry provides opportunity for various types of analysis of proteins/peptides. For example, a particular protein may be selected by one mass spectrometer from the multiple proteins present in the sample and then introduced into a second machine where it is fragmented into several peptides, which are then introduced into a third machine for the analysis of the molecular mass of the different fragments. The data obtained from such an analysis is used to decipher the entire amino acid sequence of that particular protein.

3.2.2.1.10 Amino Acid Fragmentation Pattern of a Protein. In a spectrometer, a protein is usually fragmented by breakage of the peptide bond, i.e., the bond between CO-NH groups, in the backbone of the protein giving rise to one ionized or charged fragment and another neutral fragment. When the bond between CO-NH is broken, the charge may stick either to the C- or to the N-terminal of the generated fragments. Thus, two different kinds of charged molecules are produced every time a CO-NH bond is broken. In addition to breakage between CO-NH, there could be a breakage between NH-CH and CH-CO groups as well. Here, again, two kinds of charge molecules will be produced after each breakage: one carrying the charge on the C atom of the fragment and another carrying the charge on the N atom of the fragment. Thus, altogether six kinds of fragments are produced as a result of breakage in the NH-CH, CO-NH, and CH-CO groups in the backbone of the protein. The fragments containing the charge on the N- terminal are designated as a, b, and c ions, whereas the fragments with charge on the C-terminal are called x, y, and z ions, respectively. The most commonly produced fragments are derived from the breakage of a CO-NH bond, and the two ions produced are the b and y kinds, which correspond to the charge retained on the N- or C-terminal of the fragments. The b and y ions are most commonly produced in a spectrometer as a result of fragmentation. In addition to breakage of bonds in the backbone of the protein, there could be breakage in the side chain(s) of amino acids, which will not be discussed any more in this book.

3.2.2.1.11 Certain methodologies used in conjunction to protein profiling by mass spectrometry. There are several methods to label proteins that assist their profiling by mass spectrometry. These methods involve labeling of proteins in vitro or in vivo with an isotope. Some of these techniques include Isotope Coded Affinity Tag (ICAT) and Stable Isotope Labeling with amino acids in Cell culture (SILAC). These are described below.

ICAT: This is a method to label Cysteine- containing proteins in vitro. by commercially available ICAT reagents. The proteins are first extracted from the control cells and from the diseased cells (such as cancer cells). The extracted proteins are separately and differentially labeled in their Cysteine residues by interaction with commercially available ICAT reagent; the proteins from control cells are labeled with ICAT reagent containing C12 whereas the protein from cancer cells is labeled with C13 ICAT reagent. These protein samples are mixed together and subjected to tryptic digestion and purified by Ion exchange chromatography and then the cysteine-containing peptides are isolated by affinity chromatography on avidin. The mixed sample of cysteine-containing peptides with C12 and C13 isotope labels are identified and quantified by LC/MS/MS by the shuttle differences in their mass. This method allows the simultaneous protein profiling of two different cell lines.

SILAC: This is another mass spectroscopy based method that is more commonly used to assess the abundance of proteins in two different populations of cells. This method involves growing two different populations of cells in culture media containing a particular amino acids containing different isotopes of a particular element for example C. Usually one culture is fed amino acid arginine with C12 where as the other culture is fed with same amino acid arginine containing heavy isotope C13. After certain period of growth the two population of cells are mixed and the protein is extracted and analyzed by mass spectrometry to reveal the occurrence of subset of proteins with C12 and C13 appearing as doublets on the mass spectrographs. The comparative heights of the bars in a doublet indicating protein on the spectrograph is used to determine the relative abundance of the proteins in the two cell populations. If the protein occurs in equal amount, the height of the protein bars on the spectrograph is equal. However, if the heights of protein bars differ, this indicates the unequal amount of proteins in the two populations. SILAC is extensively used to study the regulation of gene expression, interaction of proteins and proteins involved in cell signaling. The gene expression is usually examined by pulse labeling of protein in a cell population by feeding C13 arginine for a brief period of time.

3.2.3 Bottom-Up and Top-Down Mass Spectrometry

The determination of the molecular mass of the peptides generated after proteolysis is called "bottom-up" or peptide-level spectroscopy. In contrast, the measurements of intact proteins are called "top-down" or intact protein-level spectrometry. These two methods are not mutually exclusive but are complementary and essential to provide the complete picture of the amino acid sequence and many posttranslational modifications of a protein. These two terms, i.e., "bottom-up" and "top-down," are borrowed from genomics and are analogous to such terms used in the DNA sequencing during genome analysis.

The bottom-up spectrometry for peptide analysis is a traditional approach. In this method, the proteins are first digested by certain enzymes into smaller pieces consisting of 5–20 amino acids, and then their mass is determined by a spectrometer. These data are then matched with the masses of peptides of a known amino acid sequence in a databank to identify the protein. However, this approach has incomplete data to provide the complete picture of a protein for its identification with certainty. In addition, the bottom-up approach cannot provide information regarding the posttranslational modifications of a protein that is crucial for understanding its function, particularly its enzymatic role in a metabolic pathway. The top-down method of protein analysis begins with an intact protein. This protein is fragmented in a high-energy collision by blowing hot air in a spectrometer, and then it is analyzed to yield the molecular weight information of the intact protein that escaped fragmentation and those of the different fragments. The data so obtained are then matched with the known sequences of proteins and its pieces in a databank to generate the complete amino acid sequence of that protein. Such an analysis may show discrepancy in the molecular mass of a certain fragment, which usually indicates a modification of a protein. It can pinpoint the amino acid that bears the modification. For example, if there is an increase of 95 daltons in the molecular weight of a peptide, it is an indication of phosphorylation, i.e., addition of a PO_4 group that has a molecular weight of 95 [i.e., one atom of phosphorus ($P = 31$) + four atoms of oxygen ($O4 = 16 \times 4 = 64$)]. The top-down method was optimized by McLafferty at Cornell University (Han et al. 2006).

3.2.3.1 Bioinformatics and Determination of the Amino Acid Sequences of a Protein.

Based on the determination of molecular weights of a peptide and their pieces generated by fragmentation in a mass spectrometer, it is possible to predict the amino acid sequence of that peptide. The information regarding this peptide is then fed into the protein databank to find the protein containing the amino acid of this peptide. In this way, the entire amino acid sequence of a particular protein is usually determined with the help of several software programs and the information in the protein databank. The protein databank is based on the nucleotide sequence of individually cloned genes and those known from the genome projects. This method of determining the structure of a protein based on the analysis of mass spectrometric data with help of bioinformatics and genomics is discussed in Chapter 4.

3.3 DETERMINATION OF THE 3D STRUCTURE OF A PROTEIN

It is possible to determine the amino acid sequence of a protein and the 3D structure of the protein by methods other than mass spectrometric analysis. The amino acid sequence is determined by Edman degradation before the coming of spectrometric analysis of protein. However, the method of Edman degradation is slow and laborious and has been completely replaced by the mass spectrometry. It is not possible to determine the 3D structure of a protein by the mass spectrometric analysis. However, the amino acid sequence information obtained by mass spectrometry can be used with the help of a protein databank to predict or generate the 3D structure of some proteins. The only direct way of determining the 3D structure of a protein is the use of X-ray crystallography (XRC), X-ray diffraction, or NMR methodologies. However, XRC and NMR have their own advantages and limitations as discussed in the next section.

3.3.1 X-Ray Crystallography/X-Ray Diffraction

In this method, the crystals of a pure protein are exposed to an X-ray beam. The X-ray is diffracted by the lattice of atoms present in a protein crystal. The diffraction pattern of the X-ray depends on the number of electrons in the atoms and an the organization of atoms in a protein molecule. The diffracted X-rays detected as reflections on a detector produce a map of electrons called an electron density map. This map is used to generate the picture of atoms in a protein molecule to provide the 3D structure of that protein.

X-ray crystallography has been used to elucidate the 3D structure of macromolecules, including proteins and nucleic acids and has ushered in a new era in biology. X-ray crystallography developed almost 70 years ago by J. D. Bernal at Cambridge University, who obtained the first 3D structure of a small protein pepsin. Soon thereafter, with the use of X-ray crystallography, Max Perutz, who was a student of Bernal and Sir John Kendrew, determined the 3D structure of the first large proteins (hemoglobin and myoglobin) in 1958 for which they shared a Nobel Prize in Chemistry. The structure of the DNA double helix was also solved using X-ray defraction, which was performed by Rosalind Franklin, who was a student of Bernal working in the laboratory of Maurice Wilkins. Their conclusions regarding the double-helical structure of the DNA molecule was based on the X-ray diffraction pattern developed by Rosalind Franklin. Maurice Wilkins shared the Nobel Prize in Medicine with James Watson and Francis Crick in 1962 for the elucidation of the double-helical structure of DNA.

X-ray crystallography, as a method to generate the 3D structure of a protein, is a laborious process. One major difficulty of this methodology lies in getting the protein in pure form and then in generating the crystals of protein molecules. Proteins are difficult to obtain in crystal forms. It is more of an art than a science to produce the crystals of proteins. Proteins that are present in low abundance, and particularly the hydrophobic proteins or proteins with hydrophobic segments such as the membrane proteins, are difficult to crystallize. These proteins are not amenable to analysis by X-ray crystallography. However, some problems of protein crystallization have been overcome by recent advances, including the development of robots or crystal workstations, which use many parameters simultaneously to design the successful crystallization of protein molecules. In addition to the difficulty of crystallization, X-ray crystallography generates a tremendous amount of data that must be analyzed to generate the 3D picture of the protein molecule. Without the use of computers and bioinformatics, X-ray analysis of the crystal structure of proteins or nucleic acids could not have been possible.

3.3.2 Neutron Scattering

The 3D structure of proteins is also determined by neutron diffraction. A protein crystal is exposed to the neutron beam, and the position of atoms in a protein is determined by the scattered neutron. Unlike X-ray diffraction, the neutron is scattered by the nucleus of the atom and not by electron; therefore, this method yields a different kind of picture at the atomic level than X-ray crystallography. Only a small number of proteins have been analyzed by neutron scattering. One major problem of this method is that only a handful of neutron scattering devices are available in the world. Thus, fewer than a dozen proteins have been analyzed at the atomic level by this method.

3.3.3 Nuclear Magnetic Resonance Spectroscopy

The principle of NMR spectroscopy is based on the magnetic properties of the nuclei of certain atoms. Atoms that contain odd number(s) of protons or neutrons behave like a magnet and possess a spin. Thus, atoms of hydrogen

(H_1) or certain stable isotopes, such as H_2, C_{13}, N_{15}, P_{31}, and F_{19}, with an odd number of protons or neutrons behave like a magnet. When exposed to radio waves of a certain frequency in an external magnetic field, these atoms absorb the radio waves of the frequency equal to the frequency of their spin; this phenomenon is called "resonance." This perturbation in the state of the atom is also called "chemical shift" and is used to determine the chemical nature of the atom. After absorption of radio waves, of the atoms enter an excited stage but later emit the radiation equal to the amount radiation absorbed. The amount of emitted radiation and the time taken to emit the radiation can be measured and used to understand the nature of the atom.

NMR was developed independently by Felix Bloch at Stanford University and by Edward Purcell at the Massachusetts Institute of Technology MIT, for which they shared the Nobel Prize in Physics in 1952. NMR has a low resolution with a low signal-to-noise (S/N) ratio, even though the NMR signal provided some information about the nucleus of an atom. This difficulty of NMR was overcome by the application of Fourier transformation by Robert Ernst, and soon, 2D and multidimensional NMR became available; Robert Ernst received the Nobel Prize in Chemistry in 1991 for these developments of NMR. Later, Kurt Wuthrich developed NMR suitable for the analysis of the 3D structure of a protein molecule, for which he shared the Nobel Prize in Chemistry in 2002 (Wuthrich 2002).

Hydrogen (H_1) is the most abundant atom in a protein. NMR signals for H_1 are obtained to decipher the 3D structure of a protein. The NMR signals for H_1 vary in different positions; for example, the NMR signal for H in CH_3 is different than in CH or OH. Also, the NMR signal for H attached to C is different from the H attached to N. In addition, two adjacent H atoms bound by covalent bonds provide different NMR signals than when they exit as noncovalently bound or as single H atoms occurring alone or remote from each other. Now, it is feasible to determine the different kinds of NMR signals for H_1 because of the availability of multidimensional NMR spectroscopy. Essentially, three different kinds of NMR signals are generated. These include Nuclear Overhauser effect spectroscopy (NOESY), correlation spectroscopy (COSY), and total correlation spectroscopy (TOCSY). NMR signals from atoms that are close in space but not linked by covalent bonds are called the Nuclear Overhauser effect. This technique is useful in deciphering the positions of atoms in the structure of a protein where many distant atoms are brought together in space because of the folding of polypeptide segments, assuming the secondary and tertiary structure of the protein. COSY identifies the positions of atoms joined by chemical bonds. TOCSY identifies the atoms that are part of the networks not necessarily connected by any chemical bond. All these different measurements establish the distances among different atoms and are used to decipher the 3D structure of the protein. In addition to obtaining the information about the position of hydrogen atoms, it is necessary to obtain more information about the carbon and nitrogen atom. To do this, proteins are labeled with C_{13} and/or N_{15} using the cloned gene to express labeled proteins in bacteria. The labeled proteins are then examined by NMR to locate the position of the C and N atoms in the protein.

NMR is useful in determining the 3D structure of a protein that cannot be crystallized. NMR is usually useful for determining the structure of protein that is available in a solution, although recent advances have led to the development of solid NMR spectroscopy. NMR can determine the structure of proteins under physiological conditions. This methodology is also useful in determining the structure of proteins under denaturing conditions. It has been useful in understanding the structure of prion proteins that have an unfolded structure. An analysis of the 3D structure of unfolded proteins like prion proteins is not feasible by X-ray crystallography. Prion proteins are the cause of certain human nerve degenerative diseases. NMR is helpful in deciphering the 3D structure of small proteins of molecular weights up to 35,000 daltons; however, through recent advances in this methodology, it is possible to determine the structure of a large protein. NMR has been adapted for imaging purposes in medicine, which include revealing images of internal organs for medical diagnosis, revealing a disease, or discovering physiological alterations in human. This adapted form of NMR is called "magnetic resonance imaging" (MRI).

3.4 DETERMINATION OF THE AMOUNT OF PROTEINS

Several methods have been discussed that are used to identify proteins and determine their primary, secondary, and 3D structure to understand their roles in proteomics. However, in proteomics, it is required to know the amount of different proteins in addition to knowing their nature and structure. There are many ways to determine the amount of proteins in a sample. Most of these methods include colorimetric methods, such as the Lowry or Bradford's test and the spectrophotometric method; the latter method determines the protein content based on the amount of UV absorption at 280 nm of light. These methods determine the total amount of proteins in a sample. These methods do not provide the relative amounts of different proteins present in a sample. These methods, therefore, are of no use in the high-throughput analysis involved in proteomics. In proteomics, it is important to know the relative abundance of different proteins, because the proteome varies with the state of the cell and in response to changes in the environment. Protein contents also vary from one cell type to another in multicellular organisms or under different growth conditions of a particular cell type. In addition, normal cells may differ in protein content from the cells with a disease. Methods that can be used to quantify the different proteins in a sample are discussed next. The relative abundance of individual proteins can be determined by methods based on antigen–antibody interactions, such as radio immunoassay (RIA) in a solution, or by Western analysis, which measures the antigen–antibody interaction on a gel. The radio immunnoassay was developed by Rosalyn Yalow, who received the 1977 Nobel Prize in Medicine and Physiology for her work leading to detection of insulin in blood samples. She was the second woman in history to receive the Nobel Prize in Medicine and Physiology. Her RIA work paved the way for determinations of proteins and hormones present in small amounts in the human body. However, these methods cannot determine the relative abundance of a large number of proteins in different samples, as required in proteomics. Thus, these methods of RIA or Western analysis can provide information about the relative abundance of one protein at a time, which is contrary to the objectives of proteomics.

3.4.1 Quantitation of Proteins After Separation on a 2D Gel

The amount of proteins present in a large number band separated on 2D gel can be measured, and their relative abundance can be established by a variety of methods. First, the different samples of proteins are separated on 2D gel, and then the intensity of protein bands is measured by the intensity of dyes used to visualize the bands. The intensity of protein bands can be measured by a densitometric scan. Staining with silver stain is sensitive. Staining with certain fluorescent dye is equally useful. Alternatively, proteins in two cell types are labeled by growing cells in the presence of radioactive amino acids, such as methionine containing S_{35} sulfur. The protein samples obtained from the two cell types are then separated by 2D gel. The protein bands are visualized as spots on an X-ray film placed on the gel. The intensity of each spot on the film is determined by densitometry.

3.4.2 Differential Gel Analysis

In this method, protein samples from two different cell types are obtained, and one sample is stained with Cy3 flour (giving red color), whereas the protein in the other sample is stained with Cy5 flour (giving green color). The two samples are mixed and then run on a 2D gel together. After separation, protein bands are visualized. A particular band containing an equal amount of proteins with Cy3 and Cy5 flours is visualized as a yellow spot. However, if in a band there is a several-fold difference of a protein, then these proteins appear as a red or green spot. Those with excess of Cy3 flour appear red, whereas the band with an excess of Cy5 flour appears green. Thus, the bands with different colorations reflected the relative abundance of different proteins.

3.4.3 Quantitative Spectrometry of Proteins

Spectrometry has been used to compare the protein contents of different samples to measure the relative abundance of several proteins at the same time in a proteomic analysis. There are at least two ways to measure the relative abundance of proteins in two samples obtained from two different cell types. One cell type is grown with regular water containing hydrogen, whereas the other cell type is grown in the presence of water containing deuterium. Proteins are extracted from both cell types separately and then mixed and analyzed in a spectrometer. In such an analysis, each protein appears as doublet on the spectrometric graph: One represents the hydrogen-containing protein and another represents the protein-containing deuterium, because they will differ in their m/z ratio. The ratio of the height of the graph for these doublets will be a measure of their relative abundance.

The second method is also based on labeling proteins differentially with hydrogen and deuterium as described above. However, each protein sample is treated with isotope-coded affinity tags (ICATs). This strategy involves treating the protein samples with biotin derivatives of iodoacetamide, which interact with cystein residue in proteins. These proteins are then purified by affinity chromatography over a column containing streptavidin, which specifically binds with biotin. Later, these proteins are separated, mixed, and analyzed by a spectrometer. The spectrometer again yields a doublet of hydrogen- and deuterium-containing proteins. Their height in a doublet is measured to determine their relative abundance.

3.4.4 Protein Microarray

Protein microarray has been useful in examining the relative abundance of a large number of proteins from two different cell types or from a cell type grown under two different condition. The ability to visualize these differences in a large number of proteins from two cell types at the same time makes this method ideally suitable for proteomic analyses.

Microarray consists of a microscope glass slide exposed to a protein sample on which many antibodies are individually placed at fixed locations. This slide is then exposed to a protein solution. The spots showing protein–protein interactions are detected and used to identify a large number of proteins and their amount on a single slide. Comparisons of the intensity of spots on different slides exposed to protein solutions from different cell types or from a cell type grown under different conditions reveal the relative abundance of individual proteins. It may show an increase in the activity of a protein from different cell types. For example, the protein solution from a cancer cell may show a decrease or increase by as many as several proteins when compared with a similar microarray slide of the protein solution from a normal cell. Likewise, a microarray of protein solutions from a cell type grown in minimal medium or in enriched medium will show a difference of many proteins or the presence or absence of certain proteins. Microarray is used not only to detect the relative abundance of certain proteins but also for determining different kinds of other interactions, such as interaction of enzymes with substrate, and for identifying other small molecules that interact with proteins or enzymes. Some of these aspects of microarray are discussed in chapter 5.

3.5 STRUCTURAL AND FUNCTIONAL PROTEOMICS

The sequence of amino acids in a protein is called its primary structure, which determines the secondary and tertiary structure. In addition, it provides the other physical and functional parameters of a protein. Some of the physical parameters that are readily deduced from the amino acid sequence include its molecular weight, its pI, and the 3D structure. The function of a protein can also be deciphered from its structure, for example, whether it functions as an enzyme or as a receptor protein, as a hormone or antibody, or as a regulatory protein. It can also tell whether it resides within a cell at different locations, is a part of the membrane, or is secreted out of the cell. The 3D structure is determined by the way the sequence of amino acids in a protein folds onto itself bringing several key amino acids together in the same neighborhood. This folding pattern determines the active site for binding with a substrate or locates sites binding with a drug or other ligands and inhibitors. In addition, it may determine other regulatory sites of an enzyme protein on the surface or in the crevices of the 3D structure of the protein. A proper folding of amino acids determines the overall charge distribution of a protein leading to a structure that is thermodynamically stable. The results of the denaturation and renaturation of proteins in an in vitro study support the finding that the amino acid sequence or primary structure of a protein carries the complete information for its structure and function. It is important to understand the rules that govern the folding of a protein into its native 3D structure.

As the amino acid sequence of a protein is encoded in the nucleotide sequence of a gene, it has become possible to determine the amino acid sequences or primary structures of large numbers of proteins by deciphering the nucleotide sequences of an organism deposited in the gene databank.

This in turn has resulted in accumulation of the amino acid sequences of a large number of proteins of an organism. This is called the PDB. The PDB of yeast carries the information about all 6000 proteins encoded by the yeast genome. The PDB of higher organisms is far from complete; the annotation of large stretches of nucleotide sequences in humans and other higher organisms remains to be deciphered into proteins with known functions. Traditionally, the function of a protein is determined by its biochemical analysis, most of the time by its ability to catalyze a biochemical reaction. This tedious job requires purifying a protein and then determining which biochemical reaction is being catalyzed. This is not possible at a large scale. Therefore, following a throughput approach in proteomics, the function of a protein is assigned by comparing the primary structure and/or the 3D structure of the protein in question with the primary structure and/or 3D structure of a large number of proteins available in the PDB. The various aspects of the relationship between protein structure and function are discussed next.

3.5.1 Moonlighting by Protein

Before we explore the relationship between the structure and function of a protein, it is pertinent to mention that usually one function is assigned to a particular protein. This view perpetuates from the earlier concept of the one-gene–one-enzyme theory by Beadle and Tatum (1941). This theory implied one function per protein or the one gene–one enzyme–one function. We have already observed that the one gene–one enzyme is not valid in higher organisms. In humans, the structures of about 50% of proteins are determined by more than one gene. Likewise, it is now established that a protein may control more than one function (Yarnell, 2003). It is now known that a large number of proteins is multifunctional. Some examples of multifunctional proteins include proteins like the tumor suppressor protein P53 and Warner syndrome protein WRN; the latter has both helicase and nuclease functions. All multifunctional proteins carry different motifs or domains, each controlling a particular function. Thus, it supports the idea of the creation of genes by exon reshuffling, as will be discussed later. In addition, other factors control the different functions of the same protein. A protein may control different functions at different locations; for example, the enzyme phosphoglucoisomerase (PGI) catalyzes a biochemical reaction in the glycolytic pathway in the interior of a cell. However, the same enzyme, when secreted to outside of the cell, works as the cytokineneuroleukin, which facilitates the maturation of B cells into antibody-producing cells. This protein also functions as a nerve growth factor. The blood-clotting enzyme thrombin also works as cytokine when bound to a particular receptor. Likewise, the function of a protein may vary at different locations within a cell. For example, the bacterial protein PutA works as an enzyme dehydrogenase when it is associated with the plasma membrane, but it functions as a transcription factor that binds with DNA and controls transcription when occurring freely within the interior of the cell. The function of a protein may vary in different cells. For example, neuropilin in endothelial cells controls the production of new blood cells but controls the direction of the neural growth in an axion. Also, the function of a protein may vary depending on whether it exists as a monomer or as a multimer within a cell; for example, the same protein in a tetrameric form functions as glyceroldehyde-3-phophate dehydrogenase participating in the glycolytic pathway, but it functions as uracil-DNA glucosylase participating in the DNA repair pathway when occurring in the monomeric form. The assumption of more than one function by a protein has been called "moonlighting." (Yarnell, 2003).

3.5.2 Determining the Primary Structure of Proteins by Different Methods

As mentioned previously, three methods can be used to determine the sequence of amino acids in a protein to establish the primary structure of the protein. These methods include (a) deciphering the amino acids sequence from the nucleotide sequence of DNA or cDNA, (b) Edman degradation, and (c) mass spectrometry. The amino acid sequence of a protein is directly read from the DNA sequence. However, in higher organisms involving the mechanisms of splicing to generate the messenger RNA (mRNA) for the making of proteins, one has to be careful to recognize the nucleotides at the junction of exon and intron. The exon–intron junctions can be identified readily by the presence of conserved nucleotides. The coding sequence of DNA is thus established by identifying and ignoring the intervening non-coding intron sequences. The continuously coding sequence of DNA for a gene is usually obtained from the nucleotide sequence of complementary DNA (cDNA), which is made by copying the mRNA for a particular protein. Therefore, the sequence data from the cDNA library of an organism or the cDNA library of a tissue of an organism are useful in deciphering the amino acid sequences of different proteins. These data are now available readily in the gene databank and the protein databank.

3.5.2.1 Determination of Amino Acid Sequence by Edman Degradation. This method is used to determine the amino acid sequence of a polypeptide as large as 50 amino acids. Usually, much smaller overlapping peptides are generated by fragmentation of a bigger peptide subjected to mild hydrolysis or cleavage by endopeptidases. These smaller fragments are sequenced, and the order of amino acids is established by identifying the overlapping amino acids present in the sequence of smaller peptide obtained by the fragmentation of a larger peptide.

It is important to determine the total amino acid composition of a peptide before the amino acid sequence analysis is undertaken. This helps to establish the final order of amino acids in the peptide. The understanding of amino acid composition also helps to resolve the situation of whether the peptide contains two adjacent glycine (mol Wt = 57 daltons) residues or just one aspargine (mol. Wt = 114 daltons). This dilemma is based on the molecular weight data alone.

The total amino acid composition is determined by a complete hydrolysis of the peptide with 6 M HCl at a high temperature. The hydrolysate is run on a paper electrophoresis to separate the amino acids. The different amino acids are visualized by staining with a dye called dinitrophenol. All amino acids produce yellow spots except proline, which appears blue. The intensity of color determines the amount of each amino acid in a particular spot. The different amino acid spots are identified by comparing their mobility with the known amino acids used as standard during electrophoresis.

The order of amino acids in a peptide is determined by Edman degradation. As mentioned above, a longer peptide is fragmented in smaller peptides containing certain overlaps. The overlapping amino acid(s) are used to decide on the order of amino acids in the peptide. For example, a peptide containing three alanine, one arginine, two glutamic acid, two lysine, one phenylalanine, and one threonine can be broken into a smaller fragment. These were shown to have the following sequences by the method of Edman degradation arranged in the overlapping manner:

```
lys glu thr ala
            ala ala ala lys
        thr ala ala
                ala lys phen glu
                        phen glu arg
```

Therefore, the sequence of these 10 amino acids in the peptide can be deduced as lys glu thr ala ala ala lys phen glu arg.

The fragmentation of a larger peptide is acheived by chemical cleavage with cyanogen bromide or enzymatically by endopeptidase, such as trypsin or chymotrypsin.

It is always useful to determine the N-terminal amino acid in the peptide even before the sequencing by Edman degradation is attempted. The N-terminal amino acid is determined by reacting the peptide with Sanger's reagent fluorodinitrobenzene (FDNB), and then the amino acids are separated on electrophoresis and the N-terminal amino acid is identified by the color from binding with Sanger's reagent. FDBN yields a yellow color. In addition to FDBN, Dansyl chloride may be used to mark the N-terminal amino acid. The presence of more than one N-terminal amino acid in such an analysis is indicative of the presence of more than one peptide in the protein, as was found in the case of insulin. At times, an examination of the sequence of fragments generated by two different enzymes can help establish the amino acid sequence of the larger fragment. For example, the tryptic digestion of a peptide was found to generate two smaller fragments with an amino acid sequence determined by Edman degradation, as follows:

Ala ala trp gly lys; thr asn val lys

However, the digestion of the same peptide is generated by a smaller peptide with the following sequence:

val lys ala ala trp.

A comparison of these three sequences establishes that the amino acid sequence of the larger peptide must be thr asn val lys ala ala trp gly lys, as shown by Stryer (1982).

3.5.2.2 Determination of Amino Acid Sequence by Mass Spectrometry. The amino acid sequence of a protein is usually determined by the measurements of the weights of the peptides and their fragments in a mass spectrometer. Spectrometric analysis is fast and occurs on a large scale, which fits with the objectives of proteomic studies. Usually, a protein separated by the 2D gel is injected into a mass spectrometer containing two quadrupole analyzers and a TOF analyzer. The ionized protein molecules are fragmented and enter the first quadrupole analyzer. A particular peptide fragment is selectively allowed to enter the second quadrupole analyzer where it is mixed with certain inert gases, such as nitrogen and argon. The gases help break the peptide linkages in the peptide fragment. Care is taken to make such breakage in the peptide bond only once in each fragment but at different locations so as to generate smaller fragments with fewer amino acids per fragment. Alternatively, it is possible to remove one amino acid at a time from the C-terminus end to generate an array of peptides that is shorter by one amino acid from the next bigger chain in the series. Each time such fragmentation occurs, it results in a charged fragment with an N-terminal end and a noncharged piece with C-terminus. The charged fragments enter the TOF analyzer where they travel to the detector. The time of their travel is dependent on their mass; the smaller fragments move faster than the larger fragments. Therefore, the time taken to reach the detector is related directly to the mass of the molecule. The time of flight is recorded on a chart on the horizontal x-x axis; the height of the bar on the vertical y-y axis indicates the intensity or amount of each fragment. A typical result concerning the mass of the different fragments from a peptide containing an eight amino acid chain is shown below and in Figure 3.3.

Molecular weights of the peptide fragments and amino acids:

1-mer	2-mer	3-mer	4-mer	5-mer	6-mer	7-mer	8-mer	a.a./peptide
147	276	432	560	697	838	943	1030	Mol.Wt/peptide

Phe	Glu	arg	gln	his	met	asp	ser	
147	129	156	128	137	131	115	87	Mol.Wt of a.a.

Length of Peptide fragments (Mol. Wt)

Phe glu arg gln his met asp ser - 8 amino acids (1030)

Phe glu arg gln his met asp - 7 amino acids (943)

Phe glu arg gln his met - 6 amino acids (828)

Phe glu arg gln his - 5 amino acids (697)

Phe glu arg gln - 4 amino acids (560)

Phe glu arg - 3 amino acids (432)

Phe glu - 2 amino acids (276)

Phe -1 amino acid (147)

Once the amino acid sequence of the peptide is determined, this sequence is searched in the protein databank to determine whether a protein containing this sequence exists. On such a search of the protein databank for this sequence Phe glu arg gln his met asp ser, it is found that it belongs to bovine ribonuclease A. This completes the identification and sequence analysis of this protein.

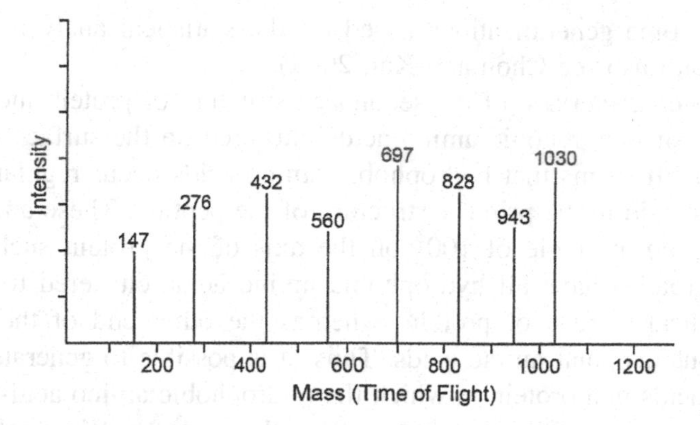

Figure 3.3: Separation and molecular weight determination of peptides by mass spectrometry. Intensity indicates the number of molecules of a peptide, whereas the time of flight indicates the molecular weight of a peptide.

3.5.3 Determining the Secondary Structure of Protein

The secondary and tertiary structure of a peptide is a function of the primary structure or the amino acid sequence of the peptide. This fact was established by Christian Anfinsen based on the denaturation or unfolding of an enzyme ribonuclease in the presence of urea and the renaturation or folding of the same enzyme after removal of the denaturing substance, i.e., urea. It is important to understand the secondary structure of a protein as a prelude to understanding of the tertiary structure and the function of proteins. It is important to know the rules that proteins follow to assume a 3D structure because of their roles in cellular function and their manipulation in biotechnology and drug design.

The linear peptide as observed in its primary structure starts folding on itself because of the interactions among the side chains of the adjacent amino acids. This leads to the formation of different structures, which include (a) helical structure called an alpha helix, (b) stranded folds called beta sheets or beta strands, and (c) random coils. Now, certain criteria can be used to predict the occurrence of these structures in the secondary structure of the protein. This was first established by Chow and Fasman (1978) based on the propensity of certain amino acids associated with these structures, i.e., helix and beta sheet. For example, the amino acids glu, met, ala, and lys are predominantly associated with the helix structure whereas the amino acids val, ile, and tyr are strongly associated with the beta sheet structure. The amino acid leucine is associated with both the helix and the beta sheet. The amino acids glycine and proline occur as breakers of the helix; proline usually occurs as the first residue in the helix. Also, asp and glu occurs at the N-terminus, whereas arg and lys occur at the C-terminus.

These are some generalizations based on the statistical analysis by Chou and Fasman (also see Chou and Kai, 2004).

Another characteristic of the secondary structure of protein includes the occurrence of hydrophobic amino acids clustered on the surface of globular proteins. It seems that hydrophobic amino acids occur regularly as the 20th amino acid in the primary structure of the peptide. These amino acids are rotated on an angle of 100° on the axis of the protein such that the globular proteins have all hydrophobic amino acids clustered to one side on the helical surface of protein, whereas the other end of the globular protein contains polar amino acids. Thus, it is possible to generate a wheel of amino acids in a protein in which the hydrophobic amino acids are clustered on one side of the helix, and the other side contains polar amino acids.

Now, several software programs can be used to examine the secondary structure of proteins in the protein databank. In addition, computer simulation is being developed to visualize the secondary and tertiary structure of proteins, which occur within microseconds of the formation of the primary structure. These computer simulations are bound to provide a better insight into the protein structure and perhaps establish the rules of secondary and tertiary structure formations.

3.5.4 Determining the Tertiary Structure of a Protein

Any mention of the structure of protein usually implies the 3D structure. Thus, the 3D structure or tertiary structure of protein has become synonymous with the word "structure" for the sake of brevity. Understanding the structure of protein is crucial for understanding its function. It is known that proteins with the same amino acid sequence may have the same structure. Thus, it is easy to figure out the structure of a protein of known sequence provided its 3D structure has been solved by a variety of methods, including X-ray crystallography or NMR.

It has also been shown that even proteins with less homology in their amino acid sequence but with the same function tend to have the same structure. Myoglobin and hemoglobin with less than 30% homology in their amino acid sequence have the same 3D structure. One of the extreme examples is plant leghemoglobin, which has no homology in amino acid sequences with hemoglobin or myoglobin. The rationale for such similarity in the structure of three different proteins such as hemoglobin, myoglobin, and plant leghemoglobin with little or no similarity in their amino acid sequence is based on the assumption that proteins with the same function must have the same structure. In the case of these three proteins, the similarity in their structure stems from the fact that they have the same function of carrying oxygen. However, an adipocyte protein AdipoQ with an unknown function was assigned the function of signaling because of its similarity in structure with a signaling protein tumor necrosis factor α (TNFα). It seems to work both ways that proteins with the same function may have the same structure or that proteins with the same structure may have the same function. Thus, one can assign the structure if one knows the function or vice versa. Therefore, the view that proteins of desperate homology but with common function must have the same structure or that proteins with a similar structure must have a common function supports the fact that structures of proteins have been more conserved than their amino acid sequences during the process of evolution and that the function of a protein is determined by its structure. It seems that proteins with little amino acid sequence homology have been selected for their function by acquiring a common structure during the process of evolution. It has been shown that these proteins with little amino acid sequence homology do possess conserved amino acids at particular places in the amino acid sequence to assume a similar 3D structure. However, there are several exceptions to this overall generalization regarding the relation between structure and function of a protein. It has been found that some protein folds may occur in different enzymes with diverse catalytic functions.

As mentioned earlier, the most common methods to determine the structure of proteins include X-ray crystallography and NMR.

Both these methods are time consuming and have their own technical difficulties. In addition, the structure of each protein must be solved individually through a laborious process. Thus, these are not high-throughput methods suitable for proteomic analysis. The computational approaches are more suitable in proteomics for determining the 3D protein structures from their primary structure, i.e., their amino acid sequence.

Several methods have been developed for this purpose. These include (a) comparative or homology modeling, (b) threading method or fold recognition, (c) ab initio method, and (d) visualization of protein folding by computer simulation.

3.5.4.1 Comparative Modeling. Comparative modeling or homology modeling is based on the fact that proteins with a homologous amino acid sequence share a common structure. A similarity of 25% or more in an amino acid sequence is considered sufficient for comparative modeling. A protein of unknown structure is assigned the structure of a known protein if both share a homology of 25% or more in their amino acid sequence. Algorithms are used to determine the amino acid sequence homology between the query and the template protein of a known structure. The homology of an amino acid sequence may be compared among a large number of proteins to determine the protein structure. Homology is examined continuously among the amino acid sequences of two or more proteins or at times discontinuously with gaps between the segments of homology as well.

3.5.4.2 Threading Method. The term "threading" was first coined by a British Chemist, Janet Thornton in 1922 (see Sternberg and Thorton 1978, Jones et al. 1992 and Jones and Hadley 2000). The method was fully described by Bowie et al. in 1991. In this method, a target or query sequence is compared against a library of templates with known structures. The protein is assigned the structure with a score of best fit. In this approach, a sequence of amino acid is threaded through the backbone structure of the template protein with the known structure and then evaluated for the score of best fit. This method has a dual approach: First, it tries to recognize a fold in the protein corresponding to a sequence in a fold library to generate a one-dimensional profile. Second, it tries to fit the protein in the 3D structure of the template assessing the interatomic distances to generate the full 3D view of the structure. This method also takes in account that in nature, there are limited numbers of protein folds because of the energetic constraints on the protein molecules. Currently, of the 4000 to 10,000 folds, only 1100 folds are known in the protein databank, and every day more and more are being added to the existing library.

3.5.4.3 Ab Initio Method. The ab initio method of prediction of the protein structure is based on thermodynamic considerations that assume that a protein must fold in a manner that the global energy requirement is minimal. This is a difficult approach, but the fact it requires only the amino acid sequence of a protein to model its structure makes it the most feasible approach. This is important particularly in view of the fact that there are more proteins for which only the amino acid sequence is known than of proteins with known structures.

This approach has been advanced by Baker et al. (2001) using an algorithm called "Rosetta." One of the major steps leading to folding into a native structure involves the interactions of certain segments with other segments in the primary structure of the protein. This results in the burial of hydrophobic regions into the interior of the folded structure and in the pairing of beta sheets. The interactions of the local segments finally lead to such a folded structure that has the minimal global energy requirement. The greatest strength of this method lies in the fact that more than 50 proteins whose structure was predicted ab initio using the Rosetta algorithm were found to have the same structure as that of proteins whose structures were solved by other physical methods, including X-ray crystallography and NMR but not known to the scientists engaged in the ab initio approach. The assumption regarding the interaction of local segments during the initiation of the folding leading to native structure is also supported during the computer simulation of folding for the purpose of visualization of such events as described below.

3.5.4.4 Visualization of Folding by Computer Simulation. Peter Kollman et al. (2000) developed the technology of molecular dynamics to visualize the folding of a small protein in a Cray supercomputer using 256 parallel processors. They used a small 36 amino acid villin headpiece subdomain viral protein and watched its folding in water for 100 μs. The unfolded structure was found to collapse within the first 20 ns which marks the beginning of folding. During the next 200 ns, the protein segments moved back and forth presumably in search of the sections with which it can interact and form a stable folded structure. These workers found brief moments of pause in between the flickers of active states finally assuming the folded structure. Similar interactions of amino acid segments have been recorded by Carol Hall in 2001 (see Nguen and Hall 2004) in the case of a polyalanine peptide.

REFERENCES

Baker, D. and A., Sali. 2001. Protein structure prediction and structural genomics. Science 294, 93–96.

Beadle, G. W. and Tatum, E. L. 1941. Genetic control of biochemical reactions in Neurospora. Proc. Natl. Acad. Sci. U.S.A. 27, 499–506.

Bowie, J. U., R. Lüthy, and D. Eisenberg. 1991. A method to identify protein sequences that fold into a known three-dimensional structure. Science 253, 164–170.

Chou, K. C. and Y. D. Kai. 2004. A novel approach to predict active sites of enzyme molecules. Proteins 55, 77–82.

Chou, P. Y. and C. D. Fasman. 1978. Empirical prediction of protein conformation. Ann. Rev. Biochem. 47, 251–278.

Comisarow, M. B. and A. G. Marshall. 1974. Fourier transformion cyclotron resonance spectroscopy. Chem. Phys. Lett. 25, 282–283.

Fenn, J. B. M. Mann, G. K. Meng, S. F. Wong, and C. M. Whitehouse. 1989. Electrospray ionization for mass spectrometryof large biomolecules. Science 246, 64.

Fenn, John. B. 2002. Electrospray wings for molecular elephants. *Les Prix Nobel. The Nobel Prizes 2002*, Editor Tore Frängsmyr, [Nobel Foundation], Stockholm, 2003.

Gorg, A. et al. 1988. The current state of two-dimensional electrophoresis with immobilized pH gradients. Electrophoresis 9: 531–546.

Gorg, A. 2000. The current state of two-dimensional electrophoresis with immobilized pH gradients. Electrophoresis 21, 1037–1053.

Han, X., M. Jin, K. Breuker, and F. W. McLafferty. 2006. Extending top-down mass spectrometry to proteins with masses >200 kDa. Science 314, 109–112.

Jones, D. T., W. R. Taylor, and J. M. Thornton. 1992. A new approach to protein fold recognition. Nature 358, 86–89.

Jones, D. T., and C. Hadley. 2000. Threading methods for protein structure prediction. In D. Higgins, and W. R. Taylor (eds.) Bioinformatics: Sequence, Structure and Databanks. Heidelberg, Germany Springer-Verlag.

Karas, M. and F. Hillenkamp. 1988. Laser desorption ionization of proteins with molecular masses exceeding 10,000 daltons. Anal. Chem. 60, 2299–2301.

Klose, J. 1975. Protein mapping by combined isoelectric focusing and electrophoresis in mouse tissues. A novel approach to testing for induced point mutations in mammals. Humangenetik 26, 231–243.

Kollman, P., I. et al.. 2000. Calculating structures and free energies of complex molecules: combining molecular mechanics and continuum models. Acc. Chem. Res. 33, 889–897.

Nguen, H. D. and K. Carol, Hall. 2004. Molecular dynamics simulation of spontaneous formation by random-coil peptides. Proc. Natl. Acad. Sci. U.S.A. 101, 16180–85.

O'Farrell, P. H. 1975. High resolution two dimensional electrophoresis of proteins. J. Biol. Chem. 250, 4007–4021.

Oppenheimer, J. A. 1986. 1945 First nuclear blast. In R. Rhodes, (ed.) The Making of the Atomic Bomb. New York: Simon & Schuster.

Stryer, L. 1982. Biochemistry Second Edition, W.H. Freeman and Company, San Francisco, CA.

Tanaka, K. 2002. The origin of macromolecule ionization by laser irradiation. Nobel Lecture 197–217.

Tanaka, K., H. Waki, Y. Ido, S. Akita, Y. Yoshida, and T. Yoshida. 1988. Protein and polymer analysis uo to m/z100.000by laser ionization time-of-flight mass spectrometry. Rapid Commun. Mass. Spectrom. 2, 151–153.

Wuthrich, K. 1986. NMR of Proteins and Nucleic Acids. New York: Wiley.

Wuthrich, K. 2002. NMR studies of structure and function of biological macromolecules. Nobel Lecture 235–267.

Yarnell, A. 2003. The double lives of enzymes. C&EN 81, 35–36.

FURTHER READING

Abersold, R. and M. Mann. 2003. Mass spectrometry based proteomics. Nature 422, 198–207.

Beavis, R. C. and Chait, B. T. 1996. Matrix-assisted laser desorption ionization mass-spectrometry of proteins. Methods Enzymol. 270, 519.

Bluegell, A. G., D. Chamrad, and H. E. Meyer. 2004. Bioinformatics in proteomics. Curr. Pharm. Biotechnol. 5, 79–88.

Borman, S. 2003. New protein fold made to order. C & ENG. 81, 11–13.

Cole, B. 2000. Some tenets pertaining to electrospray ionization mass spectrometry. J. Mass Spectrom. 35, 763–772.

Hall, D. A., J. Ptacek, and M. Snyder. 2007. Protein microarray technology. Mech. Aging Dev. 128, 161–167.

Karas, M., D. Bachmann, U. Bhar, and F. Hillenkamp. 1987. Matrix-assisted ultraviolet desorption of non-volatile compounds. Int. J. Mass Spectro. Ion Proc. 78, 53–68.

Kinter, M. and N. Sherman. 2000. Protein Sequencing and Identification Using Tandem Mass Spectrometry. New York: Wiley-Interscience.

Liebler, D. C. 2001. Intoduction to Proteomics: Tools for the New Biology. Totowa, NJ: Humana Press.

Link, A. J., J. Eng, D. M. Schieltz, E. Carmack, G. J. Mize, D. R. Morris, B. M. Garvik, and J. R. Yates. 1999. Direct analysis of protein complexes using mass spectrometry. Nature 17, 676–682.

McRee, D. E. and P. R. David. 2006. Practical Protein Crystallography. St. Louis, MO: Academic Press.

Phizicky, E., P. I. H. Bastiens, H. Zhu, M. Snyder, and S. Fields. 2003. Protein analysis on a proteomic scale. Nature. 422, 208–215.

Ramstorm, R. and J. Berquist. 2004. Miniaturized proteomics and peptidomics using capillary liquid separation and high resolution mass spectrometry. FEBS Lett. 567, 92–95.

Scapin, G. 2006. Structural biology and drug discovery. Curr. Pharm. Des. 12, 2087–2097.

Steen, H. and M. Mann. 2004. The ABC (XYZ's) of peptide sequencing. Nat. Rev. Mol. Cell. Biol. 5, 699–711.

Sternberg, M. J. E. and J. M. Thorton. 1978. Prediction of protein structure from amino acid sequence. Nature 271, 15–20.

Wagner, G., A. Kumar, and K. Wuthrich. 1981. Systematic application of twodimensional 1H nuclear—magnetic resonance techniques for studies of proteins. Eur. J. Biochem. 114, 375–384.

Xu, J., M. Li, D. Kim, and C. Dass. 2000. Principles and Practice of Biological Mass Spectrometry. New York: Wiley—Interscience.

Xu, Y. 2003. RAPTOR: Optimal protein threading by linear programming, the inaugural issue. J. Bioinform. Comput. Biol. 1, 95–117.

CHAPTER 4

PROTEOMICS OF PROTEIN MODIFICATIONS

There are many more proteins in any organism than the number of genes encoding them. In higher organisms, the one-to-one correlation between the number of genes and proteins does not exist, for example, humans have less than 25,000 genes but have up to a half million proteins. Some of this disparity in the number of genes and proteins is resolved by the splicing of transcripts, which produces many messenger RNAs (mRNA) per gene and more than one protein per gene. In addition to the mechanism of splicing, the posttranslational modification of proteins is the other major cause of the abundance of proteins in any organism. Such posttranslational modification occurs during or after the process of translation of an mRNA into proteins on the ribosomes.

The posttranslational modification of proteins is required for the specific function of the proteins, as well as their stability, degradation, and control of various biological processes. For example, certain proteins must be phosphorylated, i.e., one or more PO_4 groups are added to the protein chain. After phosphorylation, proteins become active in the signal transduction pathway, cell division, and other systems in an organism. In higher eukaryotes, one third of all proteins are phosphorylated.

Likewise, half of all proteins in humans are modified by glycosylation involving the addition of carbohydrate monomers. Certain membrane proteins are modified by attachment to fatty acids or lipids. Some other proteins like insulin or immunoglobulin are stabilized by the formation of disulfide

Introduction to Proteomics: Principles and Applications, By Nawin C. Mishra

bonds between different subunits. Many other proteins undergo degradation to become active; for example, insulin is translated from the mRNA as a preproinsulin. It is degraded by removal of terminal amino acids to yield proinsulin and degraded even more to give insulin, which then functions as a hormone to control the uptake of glucose molecules in mammalian cells. In addition, many proteins undergo ubiquitination by addition of a group of proteins called ubiquitin, which mark these proteins for disposal via hydrolysis by the protease system. Many other proteins may undergo modifications involving a change of one amino acid into another, such as the change of arginine into citrulline, the conversion of asparagine into aspartic acid, or the conversion of glutamine into glutamic acid. At times, certain proteins may undergo removal of a stretch of amino acids in the interior of protein called “intein”. The removal of intein is facilitated by a mechanism of protein splicing similar to RNA splicing in the processing of transcripts. Thus, among the different modifications that proteins undergo, most predominant are phosphorylation, glycosylation, and ubiquitination. Earlier, these modifications were discovered by analyzing one protein at a time, but now with the availability of proteomic methods involving the application of mass spectrometry, such modifications can be established readily in a large number of proteins. These and certain other miscellaneous posttranscriptional modifications that bring changes in the sequence or nature of amino acids in the proteins are discussed in this section.

4.1 PHOSPHORYLATION AND PHOSPHOPROTEOMICS

This posttranscriptional modification occurs in more than 30% of proteins in mammals. This involves the addition of one or more PO_4 groups to particular amino acids in a protein. In mammalian cells, PO_4 groups are added to the amino acids threonine, serine, or tyrosine of the protein. In contrast, amino acids such as aspartic, glutamic, and histidine are phosphorylated in prokaryotes, instead of tyrosine, serine, and threonine in eukaryotes. Occasionally in both prokaryotes and eukaryotes, phosphorylation occurs in arginine, lysine, and cystein residues of the protein. The ratio of phosphorylation of the three amino acid residues in mammalian cells is 1000:100:1 for threonine, serine, and tyrosine. Proteins are phosphorylated at more than one site, and usually a mixture of phosphorylated isomers with different levels of phosphorylation exists in the cell. Phosphorylation adds a negative charge to the proteins with the addition of the PO_{4-} group.

Phosphorylation is carried by enzyme protein kinases. The phosphorylated proteins may undergo the process of dephosphorylation by removal of phosphate group(s). The dephosphorylation is catalyzed by another enzyme called "protein phosphatase". In humans approximately 2% of the genome encodes for protein kinases and protein phosphatases. More than 500 protein kinases and more than 100 protein phosphatases have been found in humans. Yeast has approximately 120 kinase and 40 phosphatase genes. The process of phosphorylation usually activates or deactivates a protein and controls its biological or enzymatic activity. An active phosphorylated protein can become inactive after dephosphorylation. Likewise, an inactive phosphorylated protein becomes active after dephosphorylation. Proteins are phosphorylated to become active in certain pathways, such as signal transduction and cell division, as well as pathways leading to the cause of cancer in mammalian cells. That the drug Gleevec (Novartis Pharmaceuticals Corp., East Hanover, NJ), an inhibitor of BCR-Abl protein kinase, is used successfully to treat certain kinds of human cancer establishes the role of phosphorylarion in cancer. Protein kinases are considered an excellent target for treatment of cancer and other diseases resulting from the perturbation of signal transduction pathways. A concentrated effort is directed toward the development of inhibitors of different protein kinases to treat cancer and other diseases.

4.1.1 Phosphoproteomics

Phosphoproteomics is the large-scale study of proteins involved in the process of phosphorylation. Information is obtained for a large number of proteins and their different phosphorylation sites in a single attempt. This is in contrast to earlier approaches that described the biochemistry and genetics of one protein at a time. Phosphoproteomics has become possible because of the different technological advances including mass spectrometry in the analysis of proteins and the availability of the gene databank and protein databank. Phosphoproteins and their sites of phosphorylation are identified by the increase in molecular weights of the peptides obtained after analysis by mass spectrometry and the comparison of their amino acid or their inferred nucleotide sequence with the sequence information available in the protein and gene databanks.

Phophoproteomics involves several steps requiring the preparation of sample proteins, their enrichment, and tryptic digestion before analysis by the mass spectrometer.

The sample proteins are usually prepared by a several methods. Among these methods the following are commonly used: phosphoprotein isotope-coded affinity tag (PhIAT), isotope-coded affinity tag, (ICAT) and stable isotope labeling with amino acids in cell culture (SILAC). PhIAT introduces isotopes directly into phosphoserine and phosphothreonine residues of the protein.

The other two methods introduce isotopes in the protein at sites other than the phosphorylation sites. PhIAT and ICAT are used to label the proteins in vitro, whereas SILAC is used to label proteins in vivo. SILAC is useful for in vivo labeling of the proteins in cell cultures grown under different conditions that may influence the extent of phosphorylatiom. The ICAT and SILAC methods are described in Chapter 3.

Phosphoproteins are enriched on the affinity column. SILAC is usually used in conjunction with immobilized metal affinity chromatography (IMAC). The column containing the antibody against phosphotyrosine is used to enrich proteins containing phosphotyrosine. Antibodies against phosphoserine and phosphothreonine cannot be used to enrich phosphoproteins on the column because of the weak interaction with phosphoproteins. However, they are used to identify phophoproteins on the gel. All these antibodies are available commercially. The enriched phosphoproteins are subjected to digestion by trypsin; phosphorylated and unphosphorylated samples are mixed and then analyzed by a mass spectrometer as discussed in Chapter 3. Phosphoproteins are identified after mass spectrometry by the differences in the molecular weights based on the phosphate group attached to the amino acid residue of the peptides. The proteins are identified by a comparison of their amino acid sequences with the sequences present in the protein databank.

Using this protocol for the analysis of phosphoproteins, a large number of new phosphoproteins and new sites of phosphorylation have been identified in several pathways. Such an analysis by mass spectrometry has identified the several components of the signal transduction and other control systems. For examples, phosphoproteomics has provided a better insight into mating pheromome-induced signaling pathways in yeast. Phosphoproteomic methods have identified more than 500 proteins with 729 phosphorylation sites in the yeast signaling pathway that are responsive to the mating pheromone; of these, 139 phosphorylation sites were altered in response to the pheromone in yeast. In addition, this study has provided insight in to the role of phosphorylation in the RNA-processing and transport pathways that control the mRNA metabolism in yeast. In mammals, the role of phosphorylation on the components of the epidermal growth factor (EGF) receptor and their role in mRNA metabolism have been elucidated by the phosphoproteomic approach. These studies have also provided a better insight into the working of extracellular regulated kinase (ERK) protein kinase signaling pathways. Several novel phosphorylation sites in tumor suppressor proteins (TSC1 and TSC2) have been identified. Thus, it is obvious that phophoproteomics has shed much light on the role of phosphorylation on the signaling pathway by elucidating the components of this system. The study of phosphorylation is particularly important because phosphorylation sites are conserved across the species.

4.2 GLYCOSYLATION AND GLYCOPROTEOMICS

4.2.1 Glycosylation

Glycosylation is an important posttranslational modification of proteins. It involves the addition of carbohydrate moiety into proteins. The glycosylated proteins are called glycoproteins. In mammalian cells, 50% of all proteins are glycoproteins. They control major functions of proteins in the cell including the stability, anchorage, cell-to-cell interactions, antigenic and immunological specificity, secretion, reproduction, protection from pathogenesis, and almost all other functions as described in the next section.

These glycosylated proteins comprise two major components: carbohydrates and proteins. The amount of carbohydrate component may vary tremendously from 2% to 80% of the total mass. Glycosylated proteins with carbohydrates as the major component are also called "proteoglycans," as opposed to glycoproteins that contain a small proportion of carbohydrates. However, all glycosylated proteins are called "glycoproteins".

Glycoproteins contain carbohydrate moieties covalently linked to aspargine, serine, and threonine. There are two kinds of linkages of carbohydrates. In one group, carbohydrates are linked to the amino group of asparagines; these are called "N-linked glycoproteins". In the second kind of linkage, carbohydrates are linked to the OH group of serine or threonine and are called "O-linked glycoproteins".

In the N-linked glycoproteins, acetylglucosamine is attached to the N atom of the NH_2 group in asparagine of the protein. In O-linked glycoproteins, galactose or glucosyl-galactose is attached to the O atom of serine or threonine. At times in the O-linked glycoproteins, other sugar molecules, such as arabinose or mannose, may attach to serine, threonine, or proline in the protein. Several other carbohydrates may be attached to proteins in O-linked glycoproteins. A glycoprotein may contain exclusively one kind of linkage of sugar molecules or both O-linked and N-linked sugar molecules. Glycophorin containing 15 O-linked and one N-linked sugar molecules is a good example of glycoprotein containing both linkages in a protein. Glycophorin occurs in the cell membrane of erythrocytes. The properties of glycoprotein are determined by both the protein and carbohydrate components. The M, N, and MN blood types in humans are determined by the polymeric forms of the glycophorin gene, which codes for different amino acids at the 5 and 26 positions in the protein component of this glycoprotein. One of the direct effects of glycosylation is that it makes proteins more hydrophilic because of the presence of several OH groups on the carbohydrate components and thus increases the solubility of a protein. Glycoproteins are essential for almost all functions of cells. Some of these include roles in cell adhesion, antigenic, antibody, enzymatic, and hormonal functions; glycoproteins also play roles in reproduction and in cellular structure.

Glycoproteins facilitate the adhesion of a cell to another cell and control cell-to-cell interactions. Glycoprotein N-CAM helps nerve cells recognize themselves and bind to each other. This glycoprotein also facilitates the binding of nerve cell to muscle cell at the neuromuscular junction. Likewise, any ligand or molecule carrying fibronectin will bind to fibroblasts. Certain glycoproteins are antigens and antibodies. The antigenic property of human blood type is caused by the presence of glycoprotein in the cell membrane of the blood cell. The blood type A contains antigen A, blood type B contains antigen, B blood type AB contains both antigens A and B and blood type O carries no antigen, i.e., neither antigen A or B. In antigen, A the protein is glycosylated with N-acetyl galactosamine. In antigen, B the protein is glycosylated by galactose. Their antigenic property can be abolished or changed by removing their sugar moiety or by replacing one sugar for another on the blood cell surface. Antibody immunoglobulins are glycoproteins, and they cease to function as an antibody after removal of sugar moiety from immunoglobulins. Several groups of enzymes such as oxidoreductases, hydrolases, and transferases are glycoproteins; the latter may also function as the inhibitor of certain enzymes. Several hormones like human chorionic gonadotropin (hCG) which is found in human female urine during pregnancy, and erythropoietin, which controls the production of erythrocytes, are glycoproteins. Glycoproteins also function as carrier of certain hormones, vitamins, and cations. Several glycoproteins function during the processes of reproduction in humans. Glycoproteins increase the attraction of sperm to the egg, facilitate the access of sperm to the cervix, or control the penetrance of sperm to zona pellucida, and prevent polyspermy, i.e., fertilization of egg by more than one sperm. Many glycoproteins are involved in the structure of cells. Glycoproteins are constituents of cartilage and of synaptosomes, axons, and microsomes. In addition, blood-clotting proteins such as thrombin, prothrombin, and fibrinogen are glycoproteins. In bacterial cells, the capsules found in the S cell contain glycoproteins that cover the exterior of the cell wall, which gives the S cells a smooth appearance. Also the bacterial flagella controlling the movement of bacteria are made up of glycoproteins. In addition, glycoproteins like mucin participate in several other functions, such as protection of cells internal organs, and the skin.

4.2.2 Glycoproteomics

Glycoproteins are of great significance in terms of human health because of the different roles that these play in the structure and function of cells in different organisms. Many altered glycoproteins are markers of human diseases. Many cancers show alterations in specific glycoproteins. One classic example is prostate specific antigen (PSA). Glycoproteins serve as markers of several other cancers, including breast cancer. They also provide a basis for the management of human diseases by immunotherapy and pharmacological interventions. Glycoproteins are good targets of drugs and their interactions. Treatment with Herpecetin is an excellent example of immunotherapy for certain cancers.

Thus, understanding the nature of different glycoproteins has been of great interest. Several methods have been developed to investigate their structure, including the nature of the proteins, their glycosylation, and the position of attachment of specific carbohydrate monomers or polymers. These methods use the enrichments of glycoproteins by affinity chromatography and then understanding of the nature of proteins and conjugated carbohydrates by mass spectrometry. These methods are fast and throughput, and they provide the information about many glycoproteins within a certain timeframe. Certain membrane glycoproteins that are not readily soluble are rendered soluble by trypsin digestion, and the resulting soluble glycopeptides are enriched by affinity chromatography and analyzed by mass spectrometry. The enrichment of glycoproteins can be carried out by two different but major methods. The first method involves the use of a class of plant proteins called "lectin." Lectins bind specifically to a particular group of glycoproteins. Therefore, glycoproteins can be easily purified by affinity chromatography over a column containing a particular lectin. Lectins have also been used on microchips to visualize and identify different glycoproteins. However, not all glycoproteins can be affinity purified by lectins. Therefore, alternative methods have been developed for the enrichment of glycoproteins. One of these methods first developed by Aebersold et al. (2003) is based on hydrazine chemistry. In this method, glycoproteins are first oxidized by interaction with periodate and then conjugated to hydrazide resin in a column. The column is washed several times to remove glycoproteins that are not bound to the hydrazide column. Later, the bound glycoproteins are released from the column by enzymatic treatment, which hydolyzes the bond between the sugar moiety of glycoproteins and the NH_2 group of the hydrazide in the resin.

4.3 UBIQUITINATION AND UBIQUITINOMICS

4.3.1 Ubiquitination or Ubiquitinylation

Ubiquitin is a small but highly conserved protein found exclusively in eukaryotes. The characteristics of this protein were established in the early 1980s. For this monumental work, Aaron Ciechanover, Avram Hersko, and Irwin Rose shared the Nobel Prize in Chemistry in 2004.

Ubiquitination involves the addition of ubiquitin to a protein; this process marks a protein for proteolytic degradation by proteosomes. Ubiquitination is almost a universal modification of proteins, as all proteins are degraded at the end of their life. Proteins differ in their lifetime; some proteins have a half-life of a few seconds, whereas others have a half-life of up to several weeks. Hemoglobin has a half-life of almost 3 weeks. What determines the lifespan of a protein is not yet fully known. The N-terminal rule suggests that proteins with aspartic acid at the N-terminal are short lived, whereas the proteins with serine at the N-terminal are long lived, although it is not obvious why the occurrence of a particular amino acid at the N-terminal should control its life span. The process of ubiquitination is usually known to determine the stability of a protein. Ubiquitin, however, has other roles including that in the cell cycle, DNA repair, and transcription, in addition to the stability of a protein. Ubiquitin is a small but highly conserved protein and is 76 amino acids in length. It is so highly conserved that only three amino acid differences exist between ubiquitin in yeast and humans. Ubiquitin is attached to a protein through its lysine residue to the glycine residue at the C-terminal of a protein, which is then marked to degradation by proteosome. Once a molecule of ubiquitin is added to the protein, many more ubiquitin molecules may be added to the existing ubiquitin molecule. Such ubiquitination may result in monoubiqitination with only one molecule of ubiquitin, polyubiquitination with several molecules of ubiquitin attached in tandem to the first ubiquitin, or multiubiquitination when separate ubiquitin molecules are attached to the target proteins at different lysine residues. Monoubiquitination helps in the trafficking of proteins, whereas the polyubiquitination results in the degradation of proteins. The different forms of ubiquinated molecules may control other functions of the targeted proteins other than their stability, but these functions are poorly understood.

There are at least three classes of enzymes involved in the process of ubiquitination. These include the El activating enzyme, E2 conjugating enzyme, and E3 ligase. El activates ubiquitin in an energy-dependent manner by deriving energy from the hydrolysis of adenosine triphosphate (ATP) into adenosine diphosphate (ADP). E2 and E3 attach ubiquitin to the protein marked for degradation. Ubiqitinated protein then attaches to 26S proteosome with the help of a receptor or directly by the 19S regulatory subunit of the proteosome. The marked protein is degraded by the 20S catalytic subunit of the proteosome. In addition to these three classes of enzymes, there is a group of enzymes involved in the process of removal of ubiquitin from the proteosomes. This group of enzymes is called the deubiquitinating enzyme (DUB). In humans there are 500–600 ligases and about 70 DUB enzymes. The process of ubiquitination and the role of different enzymes in this process are elucidated completely.

4.3.2 Ubiquitinomics — Proteomics of Ubiquitin Modifications

Ubiquitin modifies proteins that are frayed, are products of mistranslation, or are chemically modified to become nonfunctional; then, these proteins are marked to be destroyed by hydrolysis on proteosomes. The proteomics of modifications by ubiquitin was first studied in yeast (Peng et al., 2003). These authors cloned the ubiquitin gene and added a 6x histidine tag at the beginning of the gene. The cloned ubiquitin genes with histidine tags were transfected into yeast cells carrying a deletion in the resident ubiquitin gene of the yeast chromosomes. These cells produced ubiquitin proteins with histidine hexamers. The ubiquitin associated with conjugating proteins and the proteins targeted for destruction were purified by affinity chromatography on a nickel column. The peptides obtained after tryptic digestion of ubiquitin attached proteins were analyzed by mass spectrometry and identified by a comparison with the protein sequence in the protein databank. Using this novel approach, Peng et al. (2003) established the presence of more than 1000 ubiquitination sites among 72 ubiquitin–protein conjugates in yeast. These workers also found changes in seven lysine residues of ubiquitin in protein conjugates, which suggests the diversification of polyubiquitin in yeast in vivo. This method of Peng et al. (2003) forms the basis of a similar approach in other organisms, including in humans. Following this approach of Peng et al., cloned ubiquitin genes with histidine tags have been introduced by transfection into human hepatocytes, and a many ubiquitin-associated proteins and particularly the conjugating and DUB enzymes have been characterized from human cells. In addition, inhibitors of proteosomes have been used to isolate and characterize the ubiquitin-associated proteins in humans after analysis by mass spectrometry. A survey of the process of ubiquitination across several species, including human, mouse, worm, fly, and yeast, suggest that the complexity of an organism is related to the number of E2, E3, and DUB enzymes. Several new enzymes such as 4 E1, 13 E2 97 E3, and 6 DUB have been discovered in mice. Defects in the process of ubiquitination have been found to cause diseases such as Alzheimer, Parkinson, autoimmunity, and cancer.

4.4 MISCELLANEOUS MODIFICATIONS OF PROTEINS

Proteins undergo several other modifications in addition to phosphorylation, glucosylation, and ubiquitination discussed above. Some of these include proteolysis, acetylation, methylation, sulfonation, frensylation, and sumoylation.

4.4.1 Proteolysis

Proteolysis makes the length of the amino acid chain in proteins shorter. In all proteins, the N-terminal amino acid methionine occurring as the initiation amino acid is removed as soon as or even before the synthesis of a protein is over. In addition, many proteins are originally synthesized as a longer chain but later are shortened by proteolytic cleavage to a much shorter active form of the protein or enzyme. The common examples of this class of proteins are insulin and zymogen. These proteins are synthesized as pre-pro-proteins and undergo proteolysis first to yield as pro-proteins and then to proteins, which are the active form of these pre-pro-proteins.

4.4.2 Methylation

Several proteins undergo methylation by the addition of a methyl CH_3 group to the lysine residue in the protein. The methylation of histones is crucial for the control of gene activity.

4.4.3 Sulfation

Some proteins, like gastrin, undergo sulfation by the addition of a sulfate group SO_4 to the tyrosine residue in the protein. Sulfation is carried out by the action of two enzymatic steps mediated by two different transferases.

4.4.4 Prenylation

Certain proteins such as Ras and transducin undergo the addition of isoprenoid groups attached to cysteine residue of the protein. Isoprenoids such as farnesyl and geranyl are 15- and 20-carbon substances, respectively.

4.4.5 Hydroxylation and Carboxylation

The hydroxylation of certain proteins occurs by the addition of the OH group to proline and lysine residues in the protein. Hydroxylation is mediated in the presence of vitamin C as a cofactor for the respective hydroxylases. Other proteins may undergo carboxylation of glutamine residue. This process requires vitamin K as a cofactor.

4.4.6 Lipidation

Certain proteins lacking a transmembrane domain usually undergo the addition of a lipid molecule to the C-terminal end. This process facilitates the anchoring of that protein to the membrane.

4.4.7 Amidation

A few hormone proteins undergo amidation or the addition of an amide group at the glycine residue at the C-terminal end of the hormone. Amidation is catalyzed in a series of enzymatic steps.

4.4.8 Sumoylation

Sumoylation involves the addition of small proteins related to ubiquitin. However, sumoylation does not mark the destruction of the protein. The process of sumoylation involves several other cellular functions such as DNA repair, chromosome maintenance, and cell cycle regulation. The process of sumoylation is poorly understood.

The miscellaneous modifications described in this section involve only a small number of proteins and have not been subjected to throughput analysis by proteomic analysis and mass spectrometry.

REFERENCES

Aebersold, R. and M. Mann. 2003. Mass Spectrometry-based Proteomics. Nature. 422, 198–207.

Peng, J., D. Schwartz, J. E., Elias, C. C., Thoreen, D., Cheng, G., Marsischky, J., Roelofs, D., Finley, and S. P. Gygi. 2003. A proteomics approach to understanding protein ubiquitination. Nat. Biotechnol. 21, 921–926.

FURTHER READING

Daub, H., K. Godl, D. Brehmer, B. Kelb, and G. Muller. 2004. Evaluation of kinase inhibitor selectivity by chemical proteomics. Assay Drug DEV. Techno. 2, 215–224.

Godi, K., J. Wissing, A. Kurtenbach, P. Habenberger, S. Blencke, H. Gutbrod, K. Salassidis, M. Stein-Gerlach, M. Cotton, and H. Daub. 2003. An efficient method to identify the cellular targets of protein kinase inhibitors. Proc. Nat. Acad. Sci. 100, 15434–15439.

Gygi, S. P. et al. 1999. Quantitative analysis of complex protein mixtures using isotope-coded affinity tags. Nat Biotechnol 17 (10), 994–999.

Hershko, A., A. Ciechanover, and I. A. Rose. 1979. Resolution of the ATP-dependent proteolytic system from reticulocytes: A component that interacts with ATP. Proc. Nat. Acad. Sci. USA 76, 3107–3110.

Jensen, O. 2004. Modification-specific proteomics: Characterization of post translational modifications by mass spectrometry. Curr. Opin. Chem. Biol. 8, 33.

Joenvaara, S., I. Ritamo, H. Peltoniemiand, and R. Renkoen. 2008. N-Glycoproteomics—An automated workflow approach. Glycobiology 18, 339–349.

Kaiser, P. and L. Huang. 2008. Global approach to understanding ubiquitination. Genome Biol. 6, 1–14.

Mumby, M. and D. Brekken. 2005. Phosphoproteomics: New insights into cellular signalling. Genome Biol. 6, 230–238.

Ong, S. E., et al. 2002. Stable isotope labeling by amino acids in cell culture, SILAC, as a simple and accurate approach to expression proteomics. *Molecular & Cell Proteomics* 1, 376–86.

Sun, B., J. A. Ranish, A. G. Utleg, J. T. White, X. Yan, B. Lin, and L. Wood 2008. Shotgun glycopeptide capture approach coupled with mass spectrometry for comprehensive glycoproteomics. Mol. Cell. Proteom. 6, 141–149.

Tao, W.-J. M. A. Quan, Gritsenko, D. G. Camp, M. E. Monreo, R. J. Moore, and R. D. Smith. 2005. Human plasmaN-Glycoprotein analysis by immunoaffinity substraction, hydrazine chemistry, and mass spectrometry. J. Proteom. Res. 4, 2070–2080.

Vesilescu, J., J. C. Smith, M. Ethier, and D. Figeys. 2005. Proteomic analysis of Ubiquitinated proteins from MCF-7 breast cancer cells by immunoaffinity purification and mass spectrometry. J. Proteome. Res. 4, 2192–2200.

Zhang, H., X-J. Li, D. B. Martin, and R. Abersold. 2003. Identification and quantification of N-linked glycoproteins using hydrazine chemistry, stable labeling and mass spectrometry. Nat. Biotechno. 21, 660–665.

CHAPTER 5

PROTEOMICS OF PROTEIN–PROTEIN INTERACTIONS/INTERACTOMES

The one-gene–one-enzyme theory generated the view that biochemical reactions are catalyzed individually by many enzymes one after another in a biochemical pathway. This view perpetuated through the development of molecular biology.

However, after the development of systems biology, this view of one enzyme catalyzing one biochemical reaction in isolation has been changed to a newer understanding that a group of proteins interact in a metabolic pathway. This new approach has led to the understanding of networks of proteins and to the concept of interactomes. Interactomes are also called complexosomes because they consist of many proteins in different metabolic pathways. Analysis of interactomes has become an essential part of the understanding of the interaction and function of proteins. A complete understanding of interactomes is essential for the understanding of all metabolic processes. In addition, a better understanding of interactomes is essential for understanding diseases and the development of drugs, because both diseases and drug development are related to changes in the metabolic pathways controlled by many proteins working together in interactomes. A near-complete understanding of interctomes in yeast has been established. Methods have been developed to decipher the interaction of proteins in vivo and in vitro. Some of these approaches are discussed in this chapter.

Introduction to Proteomics: Principles and Applications, By Nawin C. Mishra

5.1 PROTEIN – PROTEIN INTERACTIONS (PPI) IN VIVO

5.1.1 Yeast Two-Hybrid Assay for Protein–Protein Interactions

AD & BD Domains of a transcription activator

Most proteins contain more than one motif or domain that performs specific function. The DNA transcription is usually facilitated by the action of a protein called Transcription activator (TA) or Transcription factor (TF). Transcription factor contains at least two domains: One that bind with the DNA (to be transcribed) at a specific site this is called DNA–binding Domain (BD) and another domain that interacts with RNA polymerase to initiate transcription, this domain is called Activation Domain (AD). A DNA segment/gene for TA/TF must contain information to code for these domains in the protein. These two domains must be present in the cell either on the same DNA segments or in different segments in order to bring the two domains together to facilitate the transcription of a gene. GAL-4 is a transcription activator in yeast and the yeast two-hybrid system provides the means to detect the presence of the BD and AD by activation of a reporter gene such as beta-galactosidase gene. The transcription of the reporter gene, beta galactsidase in the yeast-two hybrid cell is inferred from its action in producing blue yeast colonies in presence of a chromogenic substance in the growth medium.

The yeast two-hybrid (Y2H) system was developed by Fields and Song (1989). This determines the interaction of a number of proteins to a known protein by expression of a reporter gene. The principle of this technique is based on the fact that both the DNA binding domain (BD) and the activation domain (AD) are required for the transcription of a gene. This uses the gal-4 gene system in yeast. The gal-4 gene encodes a protein that is a transcription factor for the lac gene. In the presence of a gal-4 gene product that has a DNA BD and an AD, the lac gene encodes the beta galactosidase enzyme. This enzyme catalyzes the conversion of a colorless substance into a blue product, which can be visualized in the yeast colonies growing on a plate. Thus, the activation of the lac gene indicates the proper functioning of gal-4 gene by the interaction of the two domains. If any of these domains are nonfunctional, the lac gene is not transcribed, and in the absence of the enzyme beta galactosidase, the yeast colonies appear colorless. It has been shown that as long as these two domains remain in proximity, even if they exist on two separate peptides, they can activate the transcription of the beta galactosidase gene in yeast, which leads to the change to blue color of the yeast colonies. Thus, this system provides the method to determine the interaction of two proteins: one containing the gal-4 DNA BD and the other containing the transcriptional AD. The yeast colonies appear blue if these two proteins interact to bring these two gal-4 domains in proximity. If these proteins do not interact, then the inference is that there is no transcription of galactosidase gene, and therefore, the yeast colonies remain colorless. The gal-4 DNA segment containing the BD is fused to the DNA segment encoding the protein participating in the protein–protein interaction. Likewise, the gal-4 AD DNA is fused to another DNA segment encoding the protein that is being examined for interaction with the protein encoded by the DNA segment containing the BD. These two constructs of chimeric genes containing the BD and AD are used to transfect yeast cells. The transfected yeast cells produce blue colonies if these two proteins interact to bring the BD and AD together in close proximity. The interacting proteins then activate the lac gene and produce beta galactosidase, which changes the color of yeast colonies when grown in the presence of a chromogenic substance. Fields and Song (1989) in their first creation of the two-hybrid system used DNA segments encoding two known interacting proteins (SNF1 and SNF4) fused to DNA segments encoding gal-4 BD and gal-4 AD, respectively. These plasmid constructs were transfected into yeast cells. The chimeric proteins produced in the yeast cells interacted with each other to bring the BD and AD of the gal-4 transcription factor together, which caused the transcriptional activation of the beta galactosidase gene and produced the blue colonies in the growing yeast cells. In this system, the construct containing BD was called "bait" and the construct containing the AD was called "prey". The chimeric protein that contained a peptide other than SNF4 or another suitable interacting peptide with AD could not fall prey to bait containing the SNF1 and BD segments. Thus, this system became a great tool to identify the interacting proteins with AD against a known bait containing BD by their ability to restore the transcriptional activation of the beta galactosidase gene and to produce the blue colonies in the growing yeast cells transfected with appropriate constructs in pairwise combinations.

Several advances have been made in the yeast two-hybrid system since its inception, such as the use of a reporter system other than galactosidase. Also, the yeast two-hybrid system can be made to work by fusing the alpha mating type cell with the opposite mating type cell, i.e., a yeast cell, each separately transfected with DNA segments containing ORF for interacting proteins containing BD and AD. The construction of the yeast two-hybrid system through the process of mating is considered advantageous, as this allows the opportunity for the posttranslational modifications of the proteins required for their interactions. Also, it has been possible to construct a hybrid system to examine protein interactions using *Escherichia coli* or even mammalian cells instead of yeast cells. The yeast two-hybrid system has been instrumental in establishing the network of interacting proteins and that of interactomes in yeast and other organisms. The role of different components including the BD and AD in the yeast two-hybrid system is presented in Figure 5.1 and 5.2.

5.1.2 Phage Display

This method is called "phage display" because the interacting protein is displayed or expressed on the surface of the bacterial virus. In this method, gene encoding the protein of interest is cloned near the DNA sequence that encodes the coat protein of the phage. Then, it is used to infect bacteria. After the infection of bacteria by the restructured phage genome, new virus particles are made that express the protein of interest on the surface of the coat protein. These virus particles are enriched by interacting with an antibody against the protein attached to the surface of a well; the virus particles remain attached to the well containing the antibody, whereas others are washed out. These virus particles are used to infect a new bacterial host to produce only those progeny particles that contain the protein of interest on the surface of the viral coat. The proteins are purified from the coat protein and analyzed by mass spectrometry to identify the protein matching the amino acid sequences present in the protein databank.

5.2 ANALYSIS OF PROTEIN INTERACTIONS IN VITRO

Protein–protein interactions can be detected many ways in vitro. Some of these include coimmunoprecipitation, protein pulldown assay, chemical crosslinking, fluorescence resonance energy transfer (FRET), label transfer, and tandem affinity purification (TAP). Of these, TAP is the only high-throughput method.

The coimmunoprecipitation of proteins by an antibody is the standard procedure to demonstrate the presence of interacting proteins. In this method, when a protein is precipitated by a specific antibody, the interacting protein(s) are also precipitated along with the protein precipitated by the antigen–antibody interaction; the components of interacting protein are then visualized by Western analysis within the same band on the gel after electrophoresis.

Protein pulldown works like coimmunoprecipitation except for the use of a ligand instead of an antibody to detect the interacting proteins on gel electrophoresis. The combination of the interacting proteins is observed to move much slower on the gel than the individual proteins in the interacting group. A label transfer is used to detect the proteins with a weak or transient

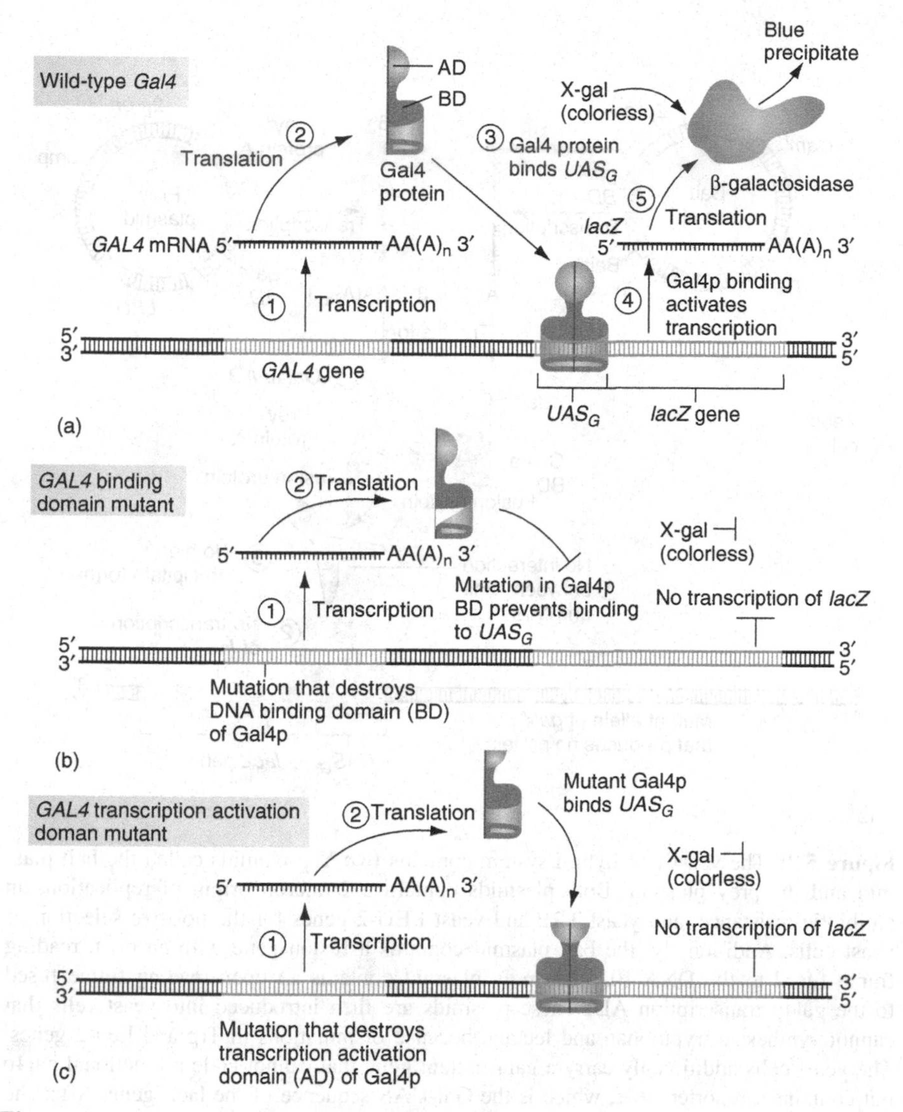

Figure 5.1: Functional assay of the two separate gal4p domains. (a) Gal4 protein produced after the transcription of the wild-type gal4 gene binds with the UASg and activates the transcription of lacZ reporter gene, which causes the expression of beta galactosidase; this hydrolyzes X-gal into a blue precipitate in the yeast cells. (b) A mutation in the beginning of of the gal4 gene destroys the DNA BD without affecting the transcriptional AD. The protein with an altered BD cannot bind UASg, which prevents it from activating the transcription of the lacZ gene. This process results in the failure of the yeast to hydrolyze X-gal and causes the yeast cell to remain colorless. (c) A mutation toward the end of the gal4 gene disrupts the AD without affecting the DNA BD. The resulting protein can bind with UASg but fails to activate the transcription of lacZ. Therefore, X-gal is not hydrolyzed and the yeast cells remain colorless. (Reproduced from *Introduction to Genetic Principles* by David Hyde, with the permission of McGraw-Hill publisher.)

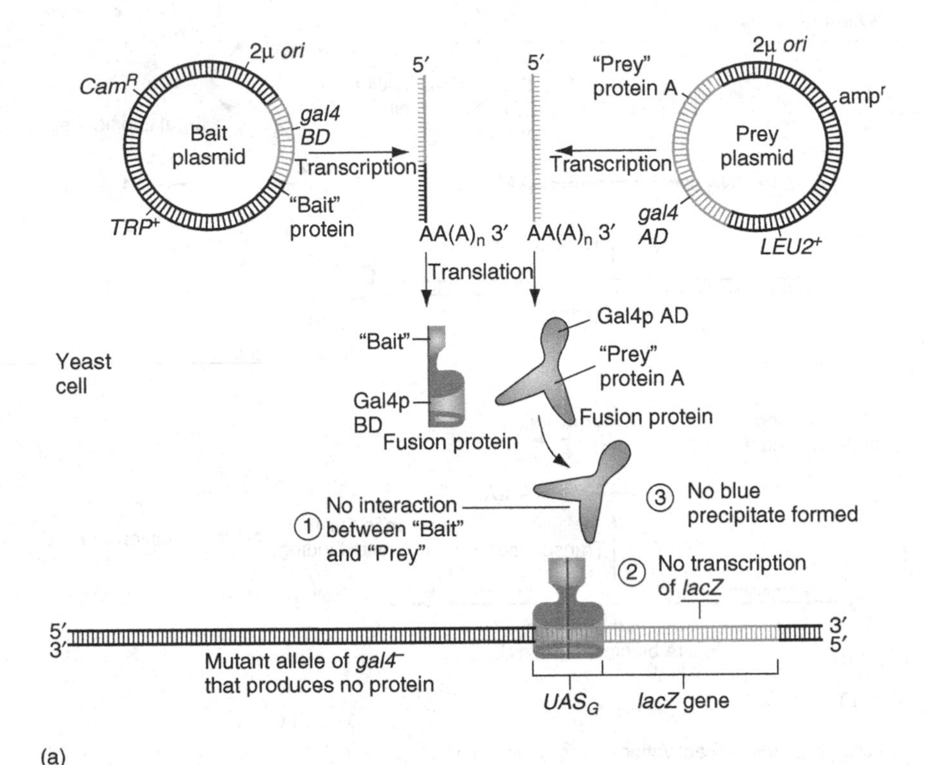

Figure 5.2: The yeast two-hybrid system contains two Yep plamids called the bait plasmid and the prey plasmid. Both plasmids contain a 2-micron origin of replication, an antibiotic resistance, and yeast TRP and yeast LEU-2 genes for the positive selection of yeast cells. Additionally, the bait plasmid contains a fusion gene with an open reading frame fused to the DNA BD. The prey plasmid contains an open reading frame fused to the gal4p transcription AD. These plasmids are then introduced into yeast cells that cannot synthesize tryptophan and leucine because of mutations in Trp and Leu-2 genes. The yeast cells additionally carry a gal4 mutant gene that cannot code a functional gal4p but contains a reporter gene, which is the Gal-UAS sequence of the lacZ gene. After the introduction of plasmids, their presence in the yeast cells is indicated by the ability of yeast cells to grow on minimal medium. (a) If the bait and prey proteins fail to produce a functional protein–protein interaction, the gal4p AD will not be brought to the UAS sequence, and the galZ gene cannot be transcribed. This causes the yeast cell to remain white even in the presence of X-gal. (b) If the bait and prey protein domains interact with eachother, then the gal4p AD is brought to the UAS element, which activates transcription of the lacZ gene. The resulting beta galactosidase protein converts the X-gal into blue precipitate. (Reproduced from *Introduction to Genetic Principles* by David Hyde, with the permission of McGraw-Hill publisher.)

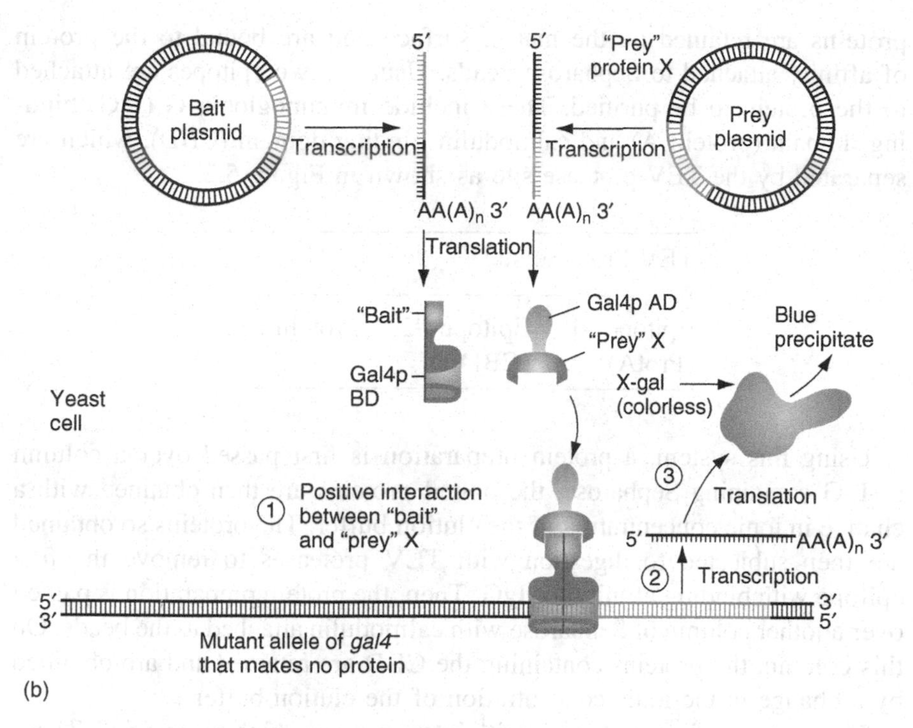

Figure 5.2: (*Continued*)

interaction by transfering a radioactive label from one protein to another protein.

TAP involves a couple of cycles of purification of interacting proteins based on their affinity to another molecule bound to a matrix and the subsequent identification of interacting proteins by mass spectrometry. This is a high-throughput method, but it cannot detect weak transient interactions because such proteins with weak interactions get separated during the steps of affinity purification. The TAP method has been used to establish the interactomes or protein networks in many organisms, including yeast. A brief description of this method is presented next.

5.2.1 TAP and Mass Spectrometry

In this method, proteins of interest are purified by use of epitope tag. The epitope tags consist of small stretch of amino acids that are fused to proteins of interest. The proteins with such epitopes are purified via column chromatography by passing an extract of proteins over a matrix such as Sepharose beads (Sigma Aldrich Co., St. Louis, MO) containing a protein with affininty for the epitope. During such column chromatography, the proteins are retained on the matrix surface and are bound to the protein of affinity attached to Sepharose beads. Usually, two epitopes are attached to the protein to be purified. These include immunoglobin G (IgG) binding domain (protein A) and calmodulin binding domain (CBD), which are separated by the TEV protease site as shown in Figure 5.2.

TEV-Protease site
N———,—,———,————C
Epitope #1 (ProtA)   Epitope #2 (CBD)   Protein #1

Using this system, a protein preparation is first passed over a column of IgG containing Sepharose; the bound proteins are then obtained with a change in ionic concentration of the elution buffer. The proteins so obtained are then subjected to digestion with TEV proteases to remove the first epitope with binding affinity for IgG. Then, the protein preparation is passed over another column of Sepharose with calmodulin attached to the beads. On this column, the proteins containing the CBD are retained and are obtained by a change in the ionic concentration of the elution buffer.

The proteins of interest along with interacting proteins are purified. These proteins are then identified by mass spectrometry. However, certain proteins such as those involved in DNA transactions, i.e., replication, repair, recombination, and transcription, are purified directly on a DNA Sepharose column without the use of epitope tags because these proteins readily bind to the DNA attached to the Sepharose beads in the matrix. The proteins bound to the matrix are removed by a change in the ionic concentration of the elution buffer. The proteins thus obtained are identified by mass spectrometry.

5.2.1.1 Mass Spectrometric Identification of Interacting Proteins. The preparation of proteins obtained after TAP is introduced into a spectrometer, and the different proteins are identified. In such an analysis, a sample of protein affinity purified using the epitope tag and the one without the epitope tag is labeled with stable isotopes and then compared for the relative abundance of different peaks in the two samples to determine the interacting proteins in a particular preparation. Thus, this is essentially a quantitative ICAT mass spectrometry described in Chapter 4. In case the proteins are purified over a DNA column without the use of an epitope tag, it is compared with a sample that cannot bind with DNA in the matrix. The samples are labeled with stable isotopes. The interacting proteins are identified by their relative abundance.

5.2.2 Functional Protein Microarray

As discussed previously, protein array consists of a glass slide on which certain antibodies are placed at known locations; the glass slide is then exposed to a protein sample. The proteins in the sample react with the antibody at the fixed position on the slide and then identified by interaction with the antibody printed on the slide in an array at fixed positions. Microarray is used to determine the relative abundance of different proteins in a cell type grown under different growth conditions or in cell types from both normal individuals and individuals with a disease grown under similar conditions. Protein array is also used to determine the interactions among proteins or between an enzyme with its substrate or inhibitor. The three kinds of protein microarrays that are in use, include analytical microarrays, functional microarrays, and reverse-phase microarrays (RPAs). The analytical microarrays are used to determine the expression level of different proteins and their relative abundance. This method is also used to determine the binding affinities of proteins. This is done by exposing a glass slide containing antibodies, aptamers, or affibodies printed on fixed positions on the glass slides.

The functional microarray typically consists of a collection of full-length functional proteins or protein domains printed on glass slides that are then exposed to a protein preparation from a cell that represents the entire proteome of that cell. This method is useful in determining protein–protein interactions. In addition, this method is useful in predicting the interaction of proteins with DNA, RNA, phospholipids, and small molecules. RPA includes glass slides on which a cellular protein preparation is fixed and then probed with a known antibody. This method helps in identifying the proteins that are altered and cannot bind with a known antibody in the proteome of diseased cell types. This method also identifies the proteins that are altered as a result of phosphorylation or other posttranslational modifications in normal and disease conditions or under growth conditions.

Using the method of the functional protein analysis, the entire protein collection of yeast, which consists of about 5800 proteins, has been cloned, overexpressed, and and placed on glass slides. These proteins on the glass slides contained GST-His-tags. These tags helped the affinity purification of these proteins on a nickel column before they were printed on the glass slides. These tags also helped in their identification by probing with fluorescent antiGST antibody. This analysis led to the identification of previously unknown calmodulin- and phospholipid-binding proteins in yeast. Such an analysis has also led to the identification of several transmembrane proteins of yeast.

5.3 ANALYSIS OF PROTEIN INTERACTIONS IN SILICO

This approach consists of the computational analyses of data available in the genome databank and protein databank. This approach is based on the comparison of the association of certain nucleic acid segments, encoded proteins, or protein domains in the databanks. The most common method is Rosetta Stone, which was developed by Marcotte et al. (1999). This approach is based on the assumption that certain interacting proteins may occur together as a fused protein called the Rosetta stone protein. The Rosetta Stone proteins are identified by analyzing the protein databank and are used for the construction of protein network or interactomes. For example, protein A and protein B may be considered as interacting proteins if they appeared as a single-fused protein in some organism after searching the protein databank. The interacting proteins are depicted in the following diagram.

Structure of Protein A	x-x-x-x-x-x-x-x-x
Structure of Protein B	o-o-o-o-o-o-o-o
Structure of Rosetta Protein/Fused A-B Protein	x-x-x-x-x-x-x-x-x-o-o-o-o-o-o-o-o

A computer program is used to search for the presence of the Rosetta protein in related organisms, and if found, the Rosetta protein is used to indicate that these proteins indeed interact with each other. Such findings of the Rosetta protein are used to construct the interaction network or interactome. Several other computational methods are used to establish the network of interacting proteins (Shoemaker and Panchenko 2007). One of the major advantages of computational methods is that they help to keep a check on the high occurrence of the false positives among the data obtained by the yeast two-hybrid analyses. Additionally, computational methods have been used to compare the interactomes of different organisms as discussed later in this chapter. Understanding comparative interactomics has led to the concept of the evolutionary conservation of protein complexes or interactomes. Thus, it is now established that the genes and proteins are conserved during the process of evolution, as well as the complexes of proteins or interactomes in nature. This concept of the evolutionary conservation of protein complexes is useful in understanding human diseases, dissecting the interactomes of simple model organisms such as yeast, and conducting a computational analysis. Thus, the use of animal systems can be avoided, and the side effects of the drug can be examined in silico. Just because of the sheer speed of the statistical methods and their ability to handle a large number of protein interactions simultaneously, these methods have emerged as the major tools for establishing the PPI maps of different organisms. Many databanks containing the PPI information are now available for many organisms, including humans. The availability of several protein databanks and software programs has added to the the use of statistical methods.

5.4 SYNTHETIC GENETIC METHODS TO DETERMINE PROTEIN INTERACTIONS

This method of establishing protein interactions is based on the principle that certain single mutation may not have lethal effects on the organism, but when two particular mutations among them are put together in an organism, the organism is disabled. This definition suggests that the proteins produced by these genes interact to control the phenotype. More than 4000 such mutations have been identified in yeast that may cause lethal effects when two particular mutations occur together in yeast cells.

5.5 INTERACTOMES

Most proteins occur as complexes; several complexes together can carry out diverse cellular functions in an organism. These functions may involve different metabolic pathways, cell-to-cell communication including signaling, several DNA transactions such as DNA replication, transcription, repair and recombination, and cell division and growth. Such a complex of proteins has been called an interactome or a complexosome.

Interactomes or PPI maps have been investigated in a variety of organisms, including viruses, bacteria, and several eukaryotes including yeast, worms, flies, mice, humans, and certain plants. Most of these organisms are model organisms with well-established genetics, biochemistry, and molecular biology. The study of interactomes has helped the assignment of gene function and their regulation. In addition, the study of PPI maps provides the basis for the understanding of the complexity of different organisms, which could not be explained by simple variation in the number of genes. It seems the complexity of an organism is dependent on the number of PPIs. A complex organism, like humans has more PPIs than yeast, flies, and worms. The study of interactomes has revealed that not only have the genes and proteins been conserved during evolution in the organisms but also the interactomes have been conserved.

The nature of interactomes has been elucidated by a variety of methods, including Y2H analyses, TPA followed by spectrometry, and synthetic genetic analysis. It seems that yeast has approximately 6000 genes that produce 6000 proteins. These proteins are organized transiently or stably into many interactomes that carry out all the cellular functions in this organism. The numbers of interactomes vary depending on the methods used to estimate them and among the findings of different workers using the same method. Most variation in their numbers is largely caused by false-positive interactions among proteins and is a result of different methods estimating different parameters. For example, the Y2H analysis estimates both transient and stable interactions among proteins, whereas TAP spectrometry determines mostly the stable protein interactions. In addition, most interactions are determined by the protein generated by cloning particular genes using polymerase chain reaction (PCR) technology. PCR may cause changes in the nature of proteins, which may lead to false-positive as well as to false-negative interactions among the proteins being estimated. Therefore, PCR may cause a large variation in the number of interactomes.

5.5.1 Prokaryotic Interactomes

In recent years, PPI maps of several bacteria and even that of certain viruses have been investigated. Among them, the interactome of *Escherichia coli* has been known to some extent. Among prokaryotes, *E. coli* is the best characterized organism from the points of biochemistry, genetics, and bacterial physiology. The early study of this organism revolutionized molecular biology beginning from the discovery of mating in *E. coli*. The study of this organism finally led to the development of the methods of molecular cloning via recombinant technology, which ushered in the era of genomics. Using the Y2H assay, approximately 4000 prey proteins have been found to interact with about 2700 bait proteins in *E. coli*. These protein–protein interactions include different metabolic pathways of the organism. Such protein–protein interactions also have been studied in *Treponema pallida*, which causes the sexually transmitted disease syphilis in humans. Treponema has one of the smallest genomes in bacteria. It cannot be cultured in vitro and is not amenable to analysis by the genetic techniques; therefore, the analysis of protein interactions by the Y2H assay has been helpful in understanding this organism. Approximately 1000 proteins have been analyzed by Y2H assays. Of these, more than 700 proteins produced more than 3600 interactions in Y2H assays. No interaction specific for causing syphilis has been identified from these studies. Several interactomes from other bacterial species have been analyzed as well. The bacterium, *Helicobacter pylori*, is a bacterium that has been studied extensively. This bacterium is found in about 50% of diseases in humans that infect the gastric layer and cause several diseases, such as peptic ulcers and even cancer. More than 260 pylori proteins participating in more than 1200 PPIs have been analyzed by Y2H assays. This corresponds to 47% of the *H. pylori* proteome. However, a databank for the PPI is now generated by statistical methods.

In addition to the bacterial interactomes, the protein–protein interactions of the bacteriophage have been well studied using Y2H assays. About 27 proteins of the bacteriophage T7 have been studied; most of these proteins are involved in the morphogenesis of the bacterial virus.

5.5.2 Eukaryotic Interactomes

The PPI maps of many eukaryotic organisms have been investigated. Several model organisms including flies yeast, fly, worms, and humans interactomes, have been mapped extensively. Some of these are described in this section.

5.5.2.1 Yeast Interactome. Yeast is a model simple eukaryotic organism. It has a well-established genetics, biochemistry and molecular biology. It is the first eukaryote for which the entire DNA sequence was deciphered and fully annotated. For these reasons, the PPI map was attempted in this organism to provide a better understanding of the network of biochemical reactions that determine all cellular activities, including the development, growth, and form of this organism as an example of the simplest eukaryote. A variety of methods has been used to establish the network of protein interactions. These methods include Y2H assays, spectrometry, bioinformatics, and genetic analysis. It is estimated that approximately 35,000 protein interactions occur in yeast cells. Only a part of these interactions are revealed by any of these methods, which is the main reason that few overlaps occur in the findings by these methods. The results of these approaches are also beset with other problems, such as false positives showing interactions that do not necessarily occur in the cell. These data may include false negatives that imply the inability of the system to reveal the interactions that actually occur in the cell. Some of these may be eliminated by certain statistical analysis of the data. The PPI map suggests the physical involvement of proteins in binary reactions between the bait protein and the prey protein as revealed by the Y2H assay or multiple interactions involving all proteins as revealed by the TAP spectrometry. Two models, the spoke model and the matrix model, are used to indicate the occurrence of the binary and multiple interactions, respectively. It is suggested that three times more binary than multiple interactions occur. The PPI map reveals that protein interactions occur in a hub or core where many proteins interact with each other. In this core, other protein interactions interconnect these centers of reactions, which suggests the structure of the cellular interaction network of proteins (Figure 5.3). It seems proteins with sequence homology in certain domains interact with each other more frequently than expected on chance alone, which has been observed in many organisms including *E. coli*, yeast, flies, worms, and humans. Proteins belonging to the Pfam group interact more frequently than expected by chance. Proteins have been classified based on the nature of domains. A domain is defined as the part of a protein with a hydrophobic core that has been conserved in the process of evolution. Protein families are created based on the nature of the domain that they share; these are listed in different databanks. The Pfam is one such databank of proteins that contains multiple domains of homology in their amino acid sequences. More than 1815 subgroups of Pfam proteins exist, which represent more than half of the total proteins in higher eukaryotes. The other protein domain families include Prosite, PRINT-S, SMART, and Prodom.

5.5.2.2 Fly Interactomes. *Drosophila melanogaster* represents the example of the first multicellular organism in which the entire genome sequence became available. It provided the first opportunity to understand multicellularity, tissue and organ development, and differentiation at the genomic and proteomic levels. This is a good model to understand human development and diseases at the simplest level of organization. The Y2H assay was used to study protein–protein interactions in Drosophila. A preliminary map of more than 4600 proteins involved in more than 4700 interaction has been obtained. The map of the Drosophila interactome also supports a picture of a network of proteins present as hub proteins and proteins interconnecting these hubs. It is estimated that about 65,000 interactions exist in the Drosophila interactome.

5.5.2.3 Worm Interactome. *Caenorhabditis elegans* provides an excellent opportunity to understand the role of interactomes in the determination of multicellularity, cell fate, and organ development during the life of a simple organism. The protein interaction is done in *C. elegans* by a combination of methodologies, including Y2H assays, TAP spectrometry, and in silico analysis. More than 3000 proteins related to multicellularity directly and indirectly have been examined. These proteins produced more than 4000 interactions involving different biological functions. These include vulval development, germline formation, pharynx function, protein degradation, and DNA damage response pathways. An extensive analysis by the Y2H assays identified 2900 nodes connected by

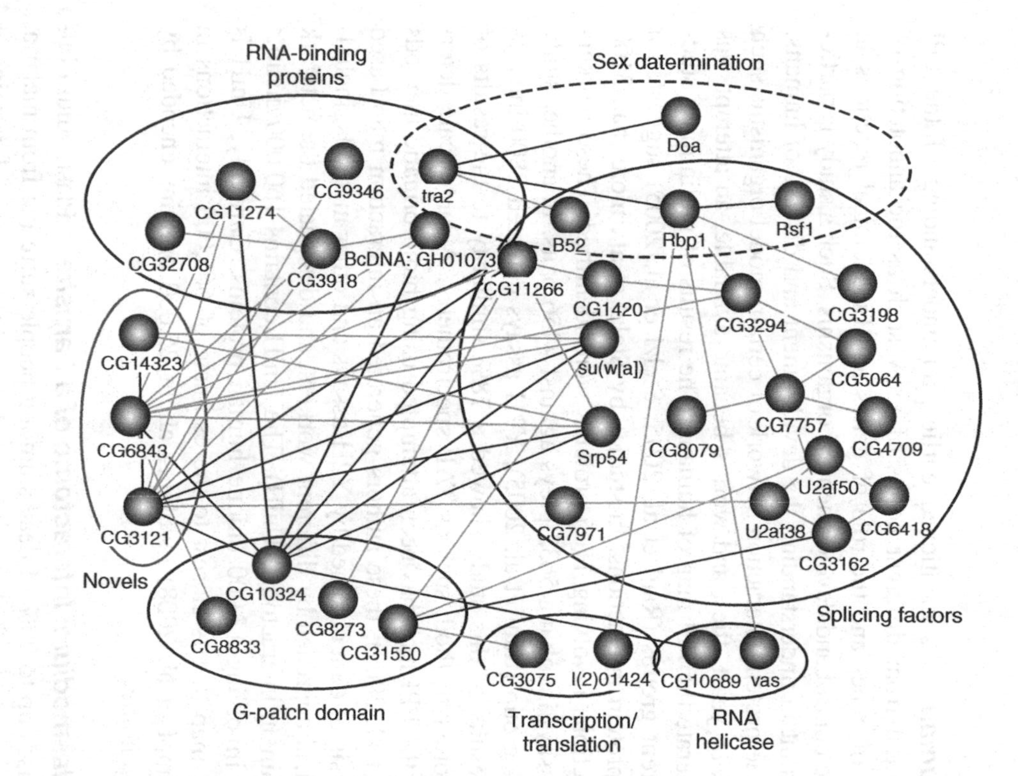

Figure 5.3: An interactome map of the proteins involved in sex determination. Such maps help in assigning functions to the genes and their involvement in the different metabolic pathways. (Reproduced from *Introduction to Genetic Principles* by David Hyde with the permission of McGraw-Hill publisher.)

about 5500 edges in the interaction network. It was revealed that the nodes comprised three classes of proteins designated as ancient, mulicellular, and worm specific. About 700 proteins belonging to an ancient class were found to have orthologs in yeast. About 1100 proteins belonging to a mulicellular class were also found in flies, Arabidopsis, and humans, whereas more than 800 proteins were worm specific (found exlusively in worms). It is also estimated that the worm interactome may consist of more than 200,000 interactions.

5.5.2.4 Human. As mentioned earlier, an understanding of the PPI map is essential from different points of view such as the annotation of the function of genes and the role of genes and biochemical reactions in the metabolic control and development of organisms. Such a study is particularly important in understanding the development and disease of humans. As soon as the protein interaction network of certain model organisms such as *Escherichia*, yeast, flies, and worms became available, an attempt was made to generate the PPI map of humans. The results of preliminary studies by different groups (Raul et al. 2005, Stelzl et al. 2005) suggested a network of protein interactions. In studies by Stelzl et al., more than 3100 novel interactions involving 1705 proteins were identified. These workers used Y2H assays involving 5632 preys against 4456 baits. In another study by the Vidal group (Raul et al. 2005), Y2H assays detected 2800 interactions. The results of this study showed a 78% similarity to the results of study by another method, such as TAP spectrometry. A large-scale determination of the PPI map has been attempted using bioinformatic methods (Chaurasia et al. 2009); these authors generated an interaction map based on the analysis of data obtained by Y2H assays and other methods, including a statistical approach by different workers, and designated the network as unified human interactome (UniH). The UniH contains 160,000 distinct interactions involving 17,000 unique human proteins. Currently, UniH is the network map that is closest to the estimate of 665,000 interactions in human (Stumpf et al. 2008) involving about 25,000 proteins encoded by the human genome.

5.5.2.5 Plasmodium: Interactome of a Parasite. Plasmodium falciparum causes up to 90% of deaths among people suffering from malaria. There has been a concerted attempt to understand the genes and proteins of this parasite through genomic and proteomic approaches. Its entire genome has been sequenced, and many genes have been annotated to assign their functions. A pentameric sequence has been identified that makes changes in the membrane of the parasite and host erythrocyte, which facilitates the entry of parasite. Most of the plasmodium proteins differed in homology from proteins of eukaryotes. About 400 proteins have been identified that target the erythrocytes; this identification corresponds to about 8% of the total genes in the plasmodium. The latest approach to understand the role of different proteins includes the generation of the PPI map with the expectation that understanding of genes and proteins in the pathogenecity of the plasmodium would provide clues and information to design antimalarial drugs. Idekar in 2005 (see Suthram et al. 2005 and Sharan et al. 2005) have shown that PPI maps differ sufficiently from those of other eukaryotes. Idekar and his group used Y2H assays to analyze more than 1300 proteins participating in more than 2800 interactions from the plasmodium. They found that the plasmodium does not share any protein interaction complex with higher eukaryotes, such as flies, worms, and humans. The plasmodium, however, showed to have three interaction complexes in common with yeast.

5.5.2.6 *Plants—Arabidopsis thaliana.* Genomes of a large number of plants are fully sequenced, but their PPI map is still not available. A start has been made in the model plant organism *Arabidopsis thaliana*, which has a small genome compared with other plants like corn and rice. This organism has been studied extensively to understand the molecular biology of plants, particularly the genetic control of flowering and nutrient requirements of plants and their adaptation to a particular environment. This plant is amenable to manipulation by several technologies of genetics and molecular biology. The Y2H assays as well as the protoplast two-hybrid system of this plant have been used to study the binary interaction among several proteins. The plant transcription factor has been well studied using these methods. Now, a database of the PPI in Arabdopsis has been based created by statistical methods.

5.5.3 Interactome During Human Development and Disease

Human development begins from an embryo, which is the result of cell proliferation and differentiation that led to the formation of more than 700 cell types containing over a trillion cells as an adult. Each cell type contains the same genome but a different proteome because of the differential expression of genes that characterize the structure and function of a particular cell type; thus, a neuron cell is much different from a muscle cell. These cell types have an entirely different protein profile that gives the neuron cell and the muscle cell their characteristic structure and function. These two cell types differ in many proteins. Several cell types predominantly produce only a few but different proteins. For example, the beta cells of the pancreas predominantly make insulin, whereas the erythrocytes make hemoglobin almost exclusively. Attempts are being made to map the interaction networks of different cell types to understand the biochemical/molecular basis of human development and differentiation. Not much progress has been made in this direction; however, a beginning is made by analyzing the protein contents of human blastocytes at different stages of development. Blastocysts grown from the discarded embryos obtained from an in vitro fertilization center in Colorado have been analyzed for their protein profile by gel electrophoresis in the laboratory of Mandy Katz-Jaffe (2006). This analysis shows differences in the content of several proteins at the early and late stages of blastocysts. Several biomarkers, such as parathyroid hormone-related peptide growth factor like epidermal growth factor, are identified as the ones responsible for early embryogenesis. Much of the difficulties of such studies lie in the scarcity of human embryos because of the ethical considerations and restrictions. It is believed that some progress can be made in this direction when it becomes possible to study stem cells. Thus, the mapping of the interactome of developing human embryo is just not possible at this time. However, it is believed that such studies in model mammals such as mice and pigs may provide the PPI network information that is relevant to human development.

5.6 EVOLUTION AND CONSERVATION OF INTERACTOMES

It seems everything in this universe has undergone evolution, which is true in the case of the living systems. It seems evolution works by the process of modifying existing systems to suit the organisms needs in a particular environment. The PPI map is no exception to this rule. The conservation of interactomes is based on the idea of the coevolution of interacting proteins. Marc Vidal first proposed that the interactomes have been conserved in the living system. However, it was Ideker who provided the evidence to support this fact. Idekar et al. (2001) compared the protein interaction networks among 6000 proteins across several species of organisms, including yeast, flies, worms, and humans. These authors showed that 71 networks were conserved across the three different species, including flies, worms, and humans. Their conclusions were based on the analysis of a large number of interactions among proteins of known and unknown functions. An important finding of their study was the assignment of function(s) to proteins that could not be predicted on the basis of their homology alone. This group also showed that *Plasmodium falciparum* (which causes malaria in humans) did not share any similarity in the interaction pathway with any higher eukaryote.

Table 5.1. Estimated number of protein interactions in different organisms.

Organism	Number of Protein Interaction
Human	650,000*
Worm	~200,000*
Fruit fly	65,000*
Yeast	35,000

*Data are based on statistical estimates by Stumpf et al. (2008).

5.7 INTERACTOMES AND THE COMPLEXITY OF ORGANISMS: IT IS THE NUMBER OF INTERACTOMES THAT MATTERS IN UNDERSTANDING THE COMPLEXITY OF AN ORGANISM AND NOT THE NUMBER OF GENES

Right from the beginning of the emergence of the chromosome theory of inheritance, geneticists have been faced with a situation called "C-value paradox." That is, the number of genes did not correlate with the complexity of an organism. For example, humans with about 24,000 genes are so different from worms with about 19,000 genes or fruit flies with 14,000 genes. Based on a recent analysis of the number of protein interactions in different organisms, Stumpf (2008) has concluded that it is the number of protein interactions that determines the complexity of an organism. For example, humans have 10 times more protein interactions than fruit flies or three times more than the worm (Table 5.1). The estimate of the number of protein interactions based on the study of Stumpf et al. is presented below. Stumpf et al. (2008) have considered only the human, fruit fly, and worm in their studies. It seems that looking at the number of protein interactions in yeast and fruit flies that it is not only the number of interactions but also the nature of interactions that must be considered when explaining the complexity of an organism. Much support for this idea regarding the role of the number of interaction in explaining the complexity of organism like humans may come from the study of plants such as corn and rice. These plants have genomes larger than humans but are far less complex than humans. Therefore, corn and rice should show a lower number of protein interaction than what is estimated for human. These data are not available currently.

5.8 INTERACTION OF PROTEINS WITH SMALL MOLECULES

Protein–protein interactions control the cellular function and structure in every organism. Most often, certain small molecules interact with proteins to facilitate their function or at times disrupt such protein interactions. Most of the molecules that facilitate the functions of proteins are cofactors, coenzymes, inhibitors, stabilizers, or allosteric effectors. A search of the protein databank revealed more than 4000 small molecules that interact with more than 20,000 proteins. Most of these are metal ions such as magnesium, calcium, and zinc, as well as various sugar molecules and nucleotides. Small molecules, in general, interact with a particular amino acid in the protein.

REFERENCES

Chaurasia, G. et al. 2009. UniHI 4: new tools for query, analysis and visualization of the human protein–protein interactome. Nucleic Acids Res. 37, D657–D660.

Fields, S., and O. K. Song. 1989. A novel genetic system to detect protein-protein interactions. Nature 340, 245–246.

Idekar, T., V. Thorsson, J. A. Ranish, R. Christmas, J. Buhler, J. M. Eng, R. Bumgarner, D. R. Goodlett, R. Abersold, and L. Hood. 2001. Integrated genomic and proteomic analyses of a systematically perturbed metabolic network. Science 292, 929–933.

Idekar, T. and A. Valencia. 2006. Bioinformatics in the human interactome project—Editorial. Bioinfoematics 22, 2973–2974.

Katz-Jaffe, M. 2006. Human embryo proteomics. Jour. Proteomic Res. 5, 1041–42.

Marcotte, E. M. et al. 1999. Detecting protein function and protein–protein interactions from genome sequences. Science, 285, 751–753.

Raul, J. F. et al. 2005. Towards a proteome-scale map of the human protein protein interaction network. Nature 437, 1173–1178.

Sharan, R., S. Suthram, R. M. Kelley, T. Kuhn, S. McCuine, P. Uetz, and T. Sittler. 2005. Conserved pattern of protein interaction in multiple species. Proc. Nat. Acad. Sci. 102, 1974–1979.

Stelz, U. et al. 2005. A human protein-protein interaction network: A resource for annotating the proteome. Cell 122, 830–2.

Suthram, S., T. Sittler, and T. Idekar. 2005. The Plasmodium protein network diverges from those of other eukaryotes. Nature 438, 108–112.

Shoemaker, B. A. and A. R. Panchenko. 2007. Deciphering protein–protein interactions. PLos Comput. Biol. 3, 42–46.

Shoemaker, B. and A. R. Panchenko. 2007. Computational method for protein interaction. PLos Comput. Biol. 3, 42–46.

Stumpf, M. P. H., T. Thorne, E. D. Silva, R. Stewart, H. J. An, M. Lappe. 2008. Estimating the size of the human interactome. Proc. Nat. Acad. Sci. 105, 6959–6964.

FURTHER READING

Arifuzzaman, M. et al. 2006. Large-scale identification of protein-protein interaction of E. ColiK12. Genome Biol. 16, 686–691.

Bader, G. D. and C. W. Hoque. 2002. Analyzing yeast protei-protein interaction data obtained from different sources. Nat. Biotechnol. 20, 991–997.

Butland, G., J. Manuel, P. Alverez, J. Li, W. Yang, V. Canadien, A. Staroateine, D. Richards, B. Beattie, N. Krogan, M. Davey, J. Parkinson, J. Greenblatt, and A. Emili. 2005. Interaction network containing conserved and essential protein complexes in E. coli. Nature 433, 531–537.

Causier, B. 2004. Studing the interactome with the yeast two-hybrid system and mass spectrometry. Mass Spectrom. Rev. 23, 350–367.

Cusick, M. E., N. Kiltgord, M. Vidal, and D. E. Hill. 2005. Interactome: Gateway into systems biology. Human Molec. Genet. 14, 171–181.

Date, S. V. and C. J. Stoeckert. 2006. Computational modeling of the Plasmodium falciparum interactome reveals protein function on a genome-wide scale. Genome Res. 16, 542–549.

Dunkley, T. P. J., S. Hester, I. P. Shadforth, J. Runions, T. Weimar, S. L. Hanton, J. L. Griffin, C. Bessant, F. Brandizzi, C. Hawes, R. B. Watson, P. Dupree, and K. S. Lilley. 2006. Mapping the Arabidopsis organelle proteome. Proc. Nat. Acad. Sci. 103, 6518–6523.

Gavin, A.-C. et al. 2002. Functional organization of the yeast proteome by systematic analysis of protein complexes. Nature 415, 141–147.

Goit, L. et al. 2003. A protein interaction map of Drosophila melanogaster. Science 302, 1727–2736.

Goll, J. and P. Uetz. 2006. The elusive yeast interactome. Genome Biol. 7, 1–8.

Han, J.-D. J., D. Dupuy, N. Bertlin, M. C. Kusick, and M. Vidal. 2005. Effect of sampling on topology predictions of protein-protein interaction network. Nature Biotechnol. 23, 839–844.

Hazbun, T. R. and S. Fields. 2001. Networking proteins in yeast. Proc. Natl. Acad. Sci. 98, 4277–4278.

Ho, Y. et al. 2002. Systematic identification of protein complexes in Saccharomyces cerevisiae by mass spectrometry. Nature 415, 180–183.

Ito, T., K. Tashiro, S. Muta, R. Ozawa, T. Chiba, and M. Nishijawa. 2000. Towards a protein-protein interaction map of the budding yeast: A comprehensive system to examine two-hybrid interactions in all possible combinations between the yeast proteins. Proc Nat. Acad Sci 97, 1143–1147.

Ito, T., T. Chiba, R. Ozawa, M. Yoshida, M. Hattori, and Y. Sakaki. 2001. A comprehensive two-hybrid analysis to explore the yeast protein interactome. Proc. Nat. Acad. Sci, 98, 4569–4574.

Jones, S. and J. M. Thorton. 1996. Princples of protein-protein interactions. Proc. Nat. Acad. Sci. 93, 13–20.

Juan, D., F. Pazos, and A. Valencia. 2008. High-confidence prediction of global interactomes based on genome-wide coevolutionary networks. Proc. Nat. Acad. Sci. 105, 934–939.

Mann, M. and O. N. Jensen. 2003. Proteomic analysis of posttranslational modifications. Nat. Biotechnol. 21, 255–261.

Li, S. et al. 2004. A map of the interactome network of the metazoan C. elegans. Science 303, 540–543.

Milstein, S. and M. Vidal. 2005. Purturbing interactios. Nat. Methods 2, 412–414.

Marti, M., R. T. Good, M. Rug, E. Knuepfer, and A. F. Cowman. 2004. Targetting Malaria virulence and remodeling proteins to the host erythrocyte. Science 306, 1930–1933.

Parks, D., S. Lee, D. Bolser, M. Schroeder, M. Lappe D. Oh, and J. Bhak. 2006. Comparative interactomics analysis of protein family interaction networks using PSIMAP(protein structural interactome map). Bioinformatics 21, 3234–3240.

Peri, S. et al. 2003. Development of humanprotein reference database as an initial platform for approaching system biology in humans. Genome Res. 13, 2363–2371.

Rain, J. C., et al. 2001. The protein-protein interaction map of Helocobacter pylori. Nature 409, 211–215.

Ramani, A. K., R. C. Bunuscu, R. J. Mooney, and E. W. Marcotte. 2005. Consolidating the set of known human protein-protein interactions in preparation for a large-scale mapping of the human interactome. Genome Biol. 6, 1–16.

Shutt, T. E. and G. S. Shadel. 2007. Expanding the mitochondrial interactome. Genome Biol. 8, 2031–2033.

Su, C., J. M. Peregrin-Alverez, G. Butland, S. Phanse, V. Fong, A. Emili, and J. Parkinson. 2008. Bacteriome.org—an integrated protein interaction data base for E.coli. Nucleic Acids Res. 36, 334–336.

Shoemaker B. A., A. R. Panchenko, S. H. Bryant. 2006. Finding biologically relevant protein domain interactions: Conserved binding mode analysis. Protein Sci. 15, 352–361.

Titz, B., S. V. Rajagopala, J. Goll, R. Hauser, M. T. Mckevitt, T. Palzkill, and P. Uetez. 2008. The binary protein interactome of Treponema pallidum—The Syphilis Spirochete. Plosone 3, 1–11.

Valente, A. X. C. N. and M. E. Cusick. 2006. Yeast protein interactome topology provides framework for coordinated-functionality. Nucleic Acids Res. 34, 2812–2819.

Walhout, A. J., R. Sordella, X. Lu, J. L. Hartley, G. F. Temple, M. A. Brasch, N. Thierry-Mieg, and M. Vidal. 2000. Protein interactions mapping in C. elegans using proteins involved in vulval development. Science 287, 116–122.

Worm, S. U. et al. 2005. A human protein-protein interaction network: A resource for annotating the proteome. Cell 122, 830–832.

CHAPTER 6

APPLICATIONS OF PROTEOMICS I: PROTEOMICS, HUMAN DISEASE, AND MEDICINE

I. PROTEOMICS, HUMAN DISEASE, AND MEDICINE

The methods of proteomics have been used extensively to understand human diseases, and the basis of medicine in alleviating these diseases as discussed in this chapter.

The fact that genes control diseases was first suggested by Gorrod (1909) in his famous writing on the inborn errors of metabolism in 1903. In this book, he documented that alcaptonuria and several other human diseases are inherited. Later, the link among genes, proteins, and diseases, as well as the possible intervention of a disease by medicine, was suggested by the one-gene–one-enzyme theory of Beadle and Tatum in 1941. This theory implied that a particular protein controlling the cellular structure or a metabolic reaction is either missing or defective in a person who has a heritable disease. The first support to this view came from the demonstration that the hemoglobin protein is altered in persons suffering from sickle cell anemia. Thereafter, many human diseases were shown to possess a particular defective protein or are missing that protein altogether. Thus, the one-gene–one-enzyme concept became the basis of medical treatment by providing insulin, which is missing in diabetic patients, or by providing the end product of a biochemical reaction not being produced by the defective enzymatic activity of a protein in the person suffering from a

Introduction to Proteomics: Principles and Applications, By Nawin C. Mishra

disease (for example, a thyroxin supplement was given to persons suffering from thyroid diseases). Soon, the one-gene–one-enzyme theory became the basis of several kinds of therapies, including the gene therapy. Gene therapy involves adding the correct form of a gene to a person with a disease, because a defective gene produces defective proteins that cannot control a biochemical reaction. Persons with severe combined immune deficiency (SCID) with defective adenosine deaminase (Ada) enzyme can be treated by adding a correct copy of the Ada gene to the SCID patient via the techniques of gene transfer (Anderson 1992). As many diseases were shown to be hereditary and many possess defective proteins, it has been shown that there are two classes of hereditary diseases, one in which one gene controls a particular disease and another in which several genes control a particular disease. These diseases are called monogenic and polygenic, respectively. With the advent of functional genomics, the nature of polygenic inheritance of a disease such as cardiovascular disease or diabetes was elucidated by the participation of several genes controlling different proteins involved in that disease. The list of all known genes controlling human diseases with their biochemical defects are complied by Victor McKusick (1966, 1998) of Johns Hopkins University; this compendium, known as Online Mendelian Inheritance of Man (OMIM), is available online. This compendium lists 1777 genes and 1284 human disorders. An understanding of proteomics is essential for medicine because of the involvements of proteins. Particularly, proteins are the cause of diseases and they can be used to diagnose prostate cancer; the level of prostate-specific antigen (PSA) is a good biomarker of the disease. In addition, proteins are used as drugs such as insulin or human growth hormone. Proteins such as protein kinase and human epidermal growth factor receptor (HER2) are targets for drugs such as Gleevec (Novartis Pharmaceuticals, East Hanover, NJ) and Herceptin, which are used to treat cancer. Also, interactions between proteins and drugs determine the side effects of a drug. In view of these facts, proteomics has a greater role in medicine. The proteomics of humans are being investigated by different laboratories under the leadership of the Human Protein Organization (HUPO): This organization is similar to the Human Genome Organization (HUGO), which oversees the efforts to understand the genomics of humans. The efforts of HUGO and HUPO are complementary in keeping track of the progress made in human genomics and proteomics. Proteomics is simple in principle but is rigorous in practice. It involves many steps, including the preparation of the sample, separation of proteins by two-dimensional (2D) gel electrophoresis, a series of chromatography, and finally identification of the proteins by mass spectrometry (MALDI-TOF-MS) and by a comparison of protein sequence in the protein databank. This process requires concerted efforts by many investigators and a large amount of funding.

This process is conducted in laboratories with the expectation that proteins could be identified as a biomarker for diagnostics and can help with drug development. To achieve this goal, proteomic methods have been applied to study the proteomes of human body parts and pathogens; some of these are discussed next.

6.1 DISEASOME

The diseasome was conceptualized first by Marc Vidal and his group at Harvard University in 2007 as a network of human diseases. It attempts to establish a link between the gene (disease genome) and the disorder (disease phenome) in humans. Many human diseases are monogenic, in which one disease is controlled by a single gene. However, in some instances, mutations at different sites within the same gene produce a disease with different phenotypes. For example, mutations in TP53 have been known to produce cancer with 11 different phenotypes. In contrast, Zellweger syndrome can result from a mutation in any of 11 different genes. To present all these situations regarding the relation between the disease genome and the disease phenome, Vidal 2007 (see Goh et al. 2007) developed the concept of diseasome in the form of a graph with genes and diseases. In this graph, the one-gene–one-disease link was represented by one line connecting the gene with the disease, whereas the other situations concerning the gene–disease link were shown by distinct lines joining a disease to several genes or, conversely, one gene to several diseases.

This conceptual scheme of the diseasome is supported by several facts, including the genetic controls of diseases, the underlying protein–protein interactions as discussed in a previous chapter, and the integrated nature of the metabolism in the cell. A view of the diseasome is presented in Figure 6.1.

Vidal and others have examined 1284 gene–disease links. They have shown that in at least 867 examples, one gene is connected to one disease. However, in 516 cases, a particular disease was controlled by more than one gene, such that several diseases shared some of the components of other diseases. Goh et al. (2007) found that complex diseases, such as cancer; cardiovascular diseases, and neurological disorders, were heterogeneous, and controlled by several genes, and shared some genes, whereas metabolic disease were, in general, controlled by single gene.

6.2 MEDICAL PROTEOMICS

Medical proteomics attempts to describe human proteins, as proteins are valuable to medicine as drugs or as target to study drugs and their side

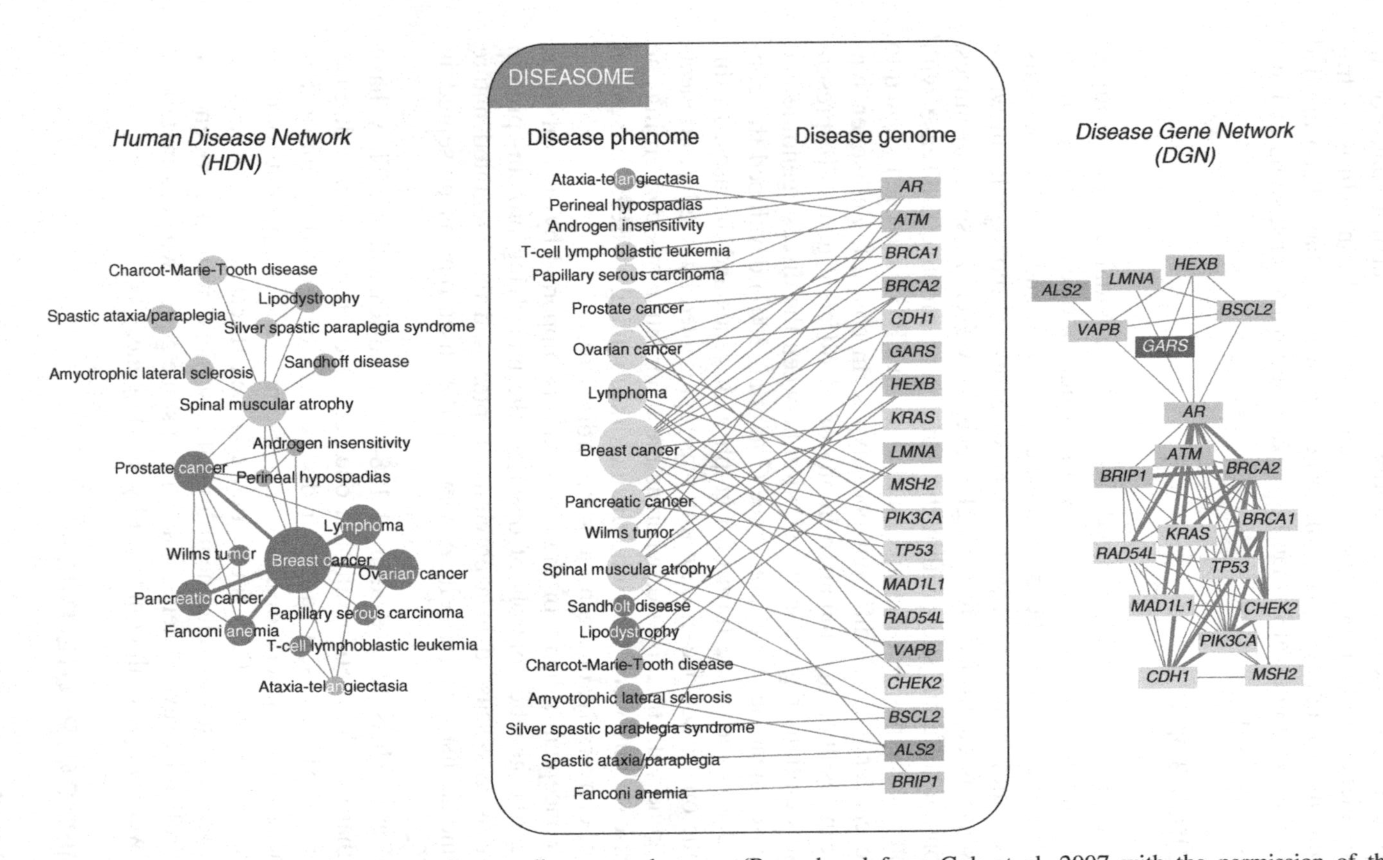

Figure 6.1: Diseasomes—The relation among human diseases and genes. (Reproduced from Goh et al. 2007 with the permission of the Proceedings of the National Academy of Sciences, USA). Center—Circles and rectangles show the relation between disorders and the genes controlling them. Left—Human Disease Network (HDN) projection of disease bipartite graph; two disorders are connected if they are controlled by the involvement of the same gene. Right—Disease Gene Network (DGN) projection; two genes are connected if they are involved in the same disorder. The size of the circles and rectangles is proportional to the number of genes and disorders involved.

effects. Thus, concerted efforts are ongoing under the leadership and sponsorship of HUPO to describe the structure and function of proteins as well as their interactions to evaluate their role in discovery of new drugs and to understand and ameliorate their serious side effects. Currently, the proteomics of different human parts are under investigation. Serum proteome is being investigated in the United States, the liver proteome is being studied in China, and the brain proteome is investigated in Germany under the sponsorship of HUPO. Different laboratories are investigating different proteomes, including the cancer proteome and cardiovascular proteome.

6.2.1 Body Fluid Proteome

Proteomics of body fluids is important because it flows through the body and comes in contact with several tissues of different organs in the body; thus, it is possible to pick up proteins that may prove to be an ideal biomarker of disease and may provide clues for drug development. Many body fluids have been studied in detail, including blood (containing plasma and serum), urine, cerebrospinal fluid, amniotic fluid, salivary gland secretion, nasal secretion, tears, and nipple aspirate. The major technical challenge in the proteomics of body fluids is the large variation in the amount of a particular protein in different individuals and among the different samples taken at different times from the same individual. Therefore, many individuals must be investigated under different times or conditions to establish a baseline amount for a particular protein.

6.2.1.1 Blood/Plasma/Serum Proteome. The human heart pumps several gallons of blood through the body. Thus, blood has the opportunity to pick up many proteins from the various parts of the human body that may provide a candidate biomarker for a specific disease. Blood is complex in its constitution. It contains a multitude of proteins in a liquid phase, and it is composed of different cells such as white blood cells, red blood cells, neutrophils, and eosinophils, as well as cellular particles such as platelets. Once the blood is centrifuged in the presence of heparin, which is an anticoagulant, the different cells and cellular particles sediment at the bottom, and the clear liquid left as supernatant is called "plasma." In the absence of an anticoagulant, proteins clot in the plasma, which can be removed by centrifugation and leaves a clear liquid containing several soluble proteins; this product is called "serum." Thus, plasma is blood without the cells, and serum is plasma without the proteins that clot, such as transferrins.

A detailed proteomic study of plasma and serum has been carried out in different laboratories primarily in the United States under the sponsorship of HUPO.

6.2.1.1.1 Plasma Proteome. The plasma proteome has been studied extensively. Plasma can be obtained readily from individuals by drawing blood using a simple procedure that is considered noninvasive. Plasma proteins have been characterized routinely by 2D gel followed by affinity chromatography, tandem mass spectroscopy, and comparison of protein sequence data in the protein databank. Some proteins such as albumin and transferrin, which are predominantly present in serum, are removed by affinity depletion chromatography to enrich the presence of other low-bundance proteins in plasma. More than 4000 proteins have been identified from human plasma, which are being developed to be used as a biomarker for several human diseases.

6.2.1.2 Salivary Proteome. In the 1950s, saliva was considered to contain only two proteins—amylase and mucin. However, it has been established that saliva contains a mixture of proteins, carbohydrates, and lipids in a physiological solution. Saliva provides immunity to several oral tissues, protects them from diseases including plaque formation, and contributes to dental development. Saliva wards off many microorganisms that can cause gum diseases and dental decay; in rare cases, these microorganisms can cause heart problems. The saliva proteome is the most non-invasive source to study proteins and to identify biomarkers for several diseases. Now, a consortium of research centers in the United States has identified 1166 proteins in human saliva; of these, 657 proteins are also found in plasma and 259 proteins are found in tears. Efforts are directed to establish changes in a particular protein by comparing the saliva proteomes from normal and diseased persons to identify a biomarker of the disease.

6.2.1.3 Others. The proteomics of human plasma has been studied extensively but is complex. There are at least two limitations in this. First, plasma contains a large number of proteins, in which it is difficult to access the relative abundance of a particular protein. Second, it is difficult to identify a biomarker in proteins of low abundance. Therefore, proteomics of certain proximal body fluids, such as urine, seminal plasma, tear fluid, and cerebrospinal fluid, have been investigated. Proximal fluids have the advantage because they come in contact with a lesser number of organs or tissue, and the amount of a particular protein is much higher. For example, urine and seminal fluid contain a larger amount of PSA than human plasma. The presence of PSA is readily detectable in urine. An extensive proteomic analysis of these proximal fluids has been performed by Anderson and Mann (2006). These workers have shown the presence of 1923 proteins in human urine, 1303 proteins in seminal fluid, and 888 proteins in human tears. Among the various proteins in proximal fluids, 190 proteins were found to exist in all three body fluids. Human urine contained 946 unique proteins not found in any other of these proximal fluids. Likewise, 352 proteins specific to seminal plasma and 178 proteins specific to tears were found. Many biomarkers were found exclusively in urine. These include corticotropin-lipotropin, which is marker for pituitary tumors, and kallikarin II, which is a marker for ovarian cancer. In addition, it included prostate secretory protein (PSP94), prostate acid phosphatase, and pancreatic secretory trypsin inhibitor, which are being evaluated as markers for prostate and pancreatic diseases.

6.2.1.3.1 Cerebrospinal Fluid. The proteome of cerebrospinal fluid can provide clues to several neurological disorders, including Alzheimer disease, Parkinson disease, and multiple sclerosis. This study has identified five proteins involved in amyloid-beta metabolism and other metabolisms. These include apolipoprotein A1, cathepsin D, and hemopexin, which were downregulated in Alzheimer patients; the transthyretin and pigment epidermal factors were elevated.

6.2.1.3.2 Amniotic Fluid. The proteome of amniotic fluid has been analyzed to reveal biomarkers for certain genetic and epigenetic diseases, preterm birth, and miscarriage. The proteomic analysis of human amniotic fluid has identified more than 800 proteins, and some of these may provide clues to different health problems of the fetus.

The effect of alcohol consumption leading to the development of fetal alcohol syndrome (FAS) has been investigated by the proteomic analysis of amniotic fluid of mice. In mice with FAS, the level of alpha fetoprotein is reduced markedly, which may be used as a biomarker of FAS.

6.2.1.3.3 Proteome of Placenta. In addition to the study of proteomics of amniotic fluid, the study of the placenta proteome has been undertaken to determine the health of the fetus and the effect of artificial fertility methods on changes in the fetus. Several proteins have been identified that may be useful in deciding the use of different artificial fertility methods.

6.2.2 Liver Proteome

The liver is an important organ from biological, physiological, pathological, and pharmacological points of view. It is second only to the brain in terms of size and complexity. It controls digestive function, formation of embryonic red blood cells, immune function, and detoxification of xenobiotics in the body. In addition, it produces retinols. It produces several proteins that bind with drugs and facilitate its distribution in the human body, which is valuable in pharmacological studies. The liver also helps to detoxify drugs by its metabolic activity and biliary secretion. From the medical point of view, the liver is subject to liver cancer, liver cirrhosis, caused by alcohol intake, and viral hepatitis, which affects millions of people worldwide. Because of its central role in biology, pathogenesis, and drug metabolism, the liver proteome has been analyzed in detail in different laboratories and mainly in China under the sponsorship of HUPO. The proteomes of normal adult, fetal, and diseased livers from different patients have been examined in detail in these studies. The HUPO liver proteome project has identified more than 5000 unique proteins. Many proteins are being evaluated as biomarkers for liver diseases. A detailed analysis of fetal liver has established 2495 distinct proteins. Proteomic analyses of fetal, adult, and cancer liver cells have been conducted as well. These studies were carried out using proteins extracted from different liver cells grown in tissue cultures. These include three different cell lines such as human fetal hepatocytes (HFH), human hepatocytes (HH4), and human liver carcinoma cells (Huh7). In these cells, a total of 2159 unique proteins were identified; of these, 496 proteins were found to exist in all three cell lines. Among these 337, 364 and 414 proteins were found characteristics of the different cell lines such as Huh7, HFH, and HH4 cells, respectively, by Yan et al. (2004). In addition, the proteomic analysis of the mouse liver has been established as a model organism. A total of 3244 unique proteins have been identified from mouse liver; 47% of these are membrane bound and about 35% have transmembrane peptides.

6.2.3 Brain Proteome

HUPO has assigned the Brain Proteome Project primarily to Germany; it is being investigated in other laboratories in other countries as well. The major objective of the brain proteome project is to decipher all the proteins in brain and to identify the proteins involved in different neurodegenerative diseases including Alzheimer disease and Parkinson disease. However, currently, not much progress has been made in the Human Brain Project except for the standardization of several protein levels, which will help identify the differences in the protein levels in the patients and establish certain biomarkers for drug development and possible treatment of the disease. In addition to Human Brain Project, the brain proteome of the mouse has been labeled as a model system. The mouse brain proteome has been characterized to a great extent. The mouse proteome has been found to contain at least 7792 proteins (Wang et al. 2007), of these 1564 proteins were identified as cysteinyl peptides. About 26% of proteins were found to be membrane proteins with a function in transportation and in cell signaling. More than 1400 proteins were found to possess transmembrane domains.

6.2.4 Heart/Cardiovascular Proteome

Proteins determine the structure and function of a cell and, thus, that of an organism. Proteins are subject to change depending on the cellular activities during the differentiation and developmental stages in the life of an organism. Proteins also change under diseased conditions as the cause or effect of the disease. Proteins also change in response to medications and other physical activities, and in response to any other changes in the environment, such as temperature, diet, and allergens. Thus, the proteome of the heart and cardiovascular system has been undertaken to identify the indicators of heart diseases and to design drugs for their treatment. A 2D gel analysis of the heart proteome has provided a protein map of the left ventricle of the human heart. This protein map identifies more than 110 unique proteins in the proteome of the left ventricle. The effect of exercise on heart proteome has been investigated in rats. The rats that underwent a regime of exercise showed alterations in 26 proteins as compared with sedentary rats. Studies of human heart and vascular system by spectrometric analysis after 2D gel or other separation of proteins have identified many proteins and alterations in these proteins under disease conditions. More than 200 proteins have been identified from the human myocardium. In addition, proteins of the sarcoplasmic/endoplasmic reticulum have been identified. Proteins such as heat shock proteins (Hsps) and mitochondrial proteins involved in energy production were reduced in their levels under condition of heart disease (Arrell et al. 2008). The levels of certain other proteins such as crystalline, HSP27, and actin were increased. In addition to changes in the level of certain proteins, under disease conditions the heart showed changes in the expression of different isoforms of the enzyme/protein or posttranslational modifications of proteins.

6.2.5 Cancer Proteome

Cancer is a devastating disease that claimed more than 7 million lives worldwide in 2005. The biology of cancer is complex because of the involvement of many genes with specific roles in the development and function of different tissues and organs. Gene expression could change because of gene mutations or changes in environmental conditions, including diet, smoking, drinking, and infections, or a combination of these factors. Because many genes and several environmental factors are involved in cancer tissues and disease, organs differ in their mechanism of the development of several types of cancers. The major aim of proteomics has been to identify a protein or a group of proteins that can serve as a biomarker for an early detection of a specific cancer before its clinical manifestation, as evidenced by biopsy and/or histological analysis of the tissue involved in a particular kind of cancer. Many biomarkers specific for a particular cancer have been reported. However, most of these are not yet applicable in clinical diagnosis or treatment of any cancer except for PSA.

The level of PSA is used to routinely monitor the possible development of prostate cancer diagnosis and management. However, the level of PSA is not 100% reliable because other factors may cause changes in its level, but an increase in its level certainly makes the attending physician cautious about the possible onset of prostate cancer. PSA has been accepted by insurance companies as a strategy to assess the risk of developing cancer. Another antigen that promises to be of use in the primary care of ovarian cancer patients is cancer antigen-125 (CA-125). Although its specificity and sensitivity is not completely reliable, other health conditions such as pancreatitis or other diseases, including liver disease or kidney disease, may influence an increase in CA-125. However, CA-125 is useful in monitoring the efficacy of a treatment regimen. It has been shown that the level of this antigen (CA-125) decreases if the line of treatment is successful. When the level of antigen remains high or increases, it is concluded that treatment is not effective and warrants a change in the treatment regimen. Another antigen that promises to be of value in the primary care of cancer patients is carcinoembryonic antigen (CEA). The level of CEA is found to increase in individuals with colorectal cancer as well as in patients with breast, lung, and pancreatic cancers. However, the level of this antigen can increase because of other conditions; for example, its level may increase by smoking. Thus, none of these antigens are reliable markers. A concept regarding the use of several markers has been developed, which is discussed later in this chapter.

6.2.6 Organelle Proteome

The study of the organelle proteome is important because certain organelles, such as mitochondria, are exclusively responsible for specific metabolic functions and their control. Mitochondria control the respiratory chain and production of energy [adenosine triphosphate (ATP)], intracellular signaling involving Ca ions, and the synthesis (heme) and degradation (urea cycle) of certain molecules in human metabolism. The study of mitochondria is important also because a defect in mitochondria may cause several human diseases, including heart diseases, Alzheimer disease, and numerous neuromuscular and neurodegerative diseases. Thus, the study of the mitochondrial proteome and that of certain other organelles is underway. The yeast mitochondrial proteome is well described, as it identified more than 1500 proteins. The study of the organelle proteome requires the isolation of mitochondria or other organelles in pure forms, and then they are analyzed by 2D gel followed by spectrometric analysis and identification of proteins. Mitochondria are unique in that they contain proteins encoded both by mitochondrial DNA and by nuclear DNA.

In human mitochondria, more than 600 proteins of 2000 proteins coded by mitochondrial and nuclear genes have been fully identified. These proteins include proteins of different pI, molecular weight, hydrophobicity, and localization, but they are all predominantly from the inner membrane of mitochondria. Many proteins were involved in energy production, signaling, biosynthesis, and ion transport. Similar amounts of proteins have been identified in the mitochondria of rats and mice by proteomic studies. In plants, the mitochondrial proteome of Arabidopsis has been well characterized.

6.2.7 Proteome of Human Parasites

In addition to infection by viruses, bacteria, and fungi, humans are infected by amoeba, protozoa, and worms, which threaten their health and cause several million deaths every year. *Plasmodium falciparum* and related species cause malaria in humans. Malaria is a devastating human disease, which causes several million deaths in developing countries. Numerous measures including the development of a vaccine have been sought for the cure and eradication of malaria. A cure through the development of drugs and vaccines has been unsuccessful because of the peculiar etiology and evasive immune system of the parasite. The parasite is unaffected by blood-stage antimalarial drugs during a certain stage in its life, such as the liver stage of the parasite. In this stage, the parasite can remain latent for years in the human liver. Also, pregnant women cannot be given these antimalarial drugs because of the possible harmful effects to the fetus. Thus, in these situations, the parasite remains unaffected by the blood-stage antimalarial drugs. Thus, the proteome analysis of the liver stage parasite has been undertaken in the plasmodium of mice. In mice, the parasites within the liver cells are detected by a cell sorter because of the of green fluorescence protein (GFP) expressed by the parasite. The proteome of the plasmodium-infected liver cell is studied by spectrometric analysis after the separation of the parasite proteins from the mice hepatocytes (Tarun et al. 2008). These authors have identified about 800 proteins that are specific to the liver stage of parasite. Some of these may be developed as a target to inhibit the fatty acid synthesis (FASII) system, which is uniquely possessed by the parasite and not by the host. In addition to plasmodium, the proteomes of several other human parasites, such as *Schistosoma mansoni* and *Toxoplasma gondii*, have been studied to find the clue to overcoming the parasitic diseases caused by them.

6.3 CLINICAL PROTEOMICS

Proteomics holds much promise for the diagnosis and treatment of human diseases and has a large role to play in the future of medicine as outlined below.

6.3.1 Genomics and Proteomics of Human Diseases

Genetics and genomics have been instrumental in determining the nature of human diseases as well as their genetic or nongenetic basis. Classic genetics and epidemiological or population genetic methods have been used to determine the genetic nature of a disease. Genetics and genomics have been used successfully to identify disease-causing genes, their chromosomal locations, and DNA sequences, and the proteins controlled by these genes have also been identified. Methods of somatic cell genetics, cytogenetics, and translocation mapping have been used to establish the chromosomal location of the disease-causing genes. Genomics has helped in determining the DNA sequence of a gene and in identifying the protein underlying these human diseases. Both the methods of classic genetics and that of genomics have been used to determine whether a disease is caused by a single gene or by multiple genes in conjunction with the environmental factors. Genomics has also made us understand why a defective gene controls a disease with a varying degree of expression (i.e., the expressivity of a gene) or why a defective gene is expressed only in a certain number of individuals (i.e., the penetrance of a gene) and not in others. This explanation has been derived by identifying the role of other genes in the human genome that modify the expression of the disease-causing genes through gene interactions. Several diseases are caused by one defective gene producing one defective protein, for example, sickle cell anemia, Huntington disease, and cystic fibrosis. However, proteomics has shown why a mutation in a single gene leads to multiple effects via a change in the protein interaction pathway or interactome, which controls a particular metabolic pathway. The multiple effects also result from a change in a particular pathway on the function of other metabolic pathways. Several diseases, such as high blood pressure, diabetes, obesity, high cholesterol, and many cardiovascular and neurological disorders are controlled by many genes in conjunction with the environmental factors. The science of proteomics is revealing the role of the multiple gene basis of diseases and the role of environmental factors by identifying the roles of different proteins and their interactions in metabolic pathways. Proteomics has also been useful in explaning why certain cancers can be controlled by certain drug(s) and not by others. Proteomics has also been useful in making us understand side effects of a particular drug. Proteomics is being used to identify certain protein(s) that may be used as an indicator or biomarker of certain diseases. Proteomics is useful to understand the biochemical basis of a disease, as well as the diagnosis and treatment of a disease.

6.3.2 Proteomics of Diagnostic Markers and Drug Development

Proteomics is crucial in understanding the cause and management of a disease because proteins can be used both as biomarker and as target of a drug. The idea that proteins can be used as biomarkers for human diseases began with the discovery of a protein by Henry Bence-Jones in 1847; later, this protein was called the Bence-Jones protein. It was identified by Kyle in 1994 as a free antibody light chain that is produced in excess by tumors; because of its smaller size, this protein can pass through the kidney and appears in the urine of cancer patients who suffer from myeloma. This protein was also found in the serum of the myeloma patients. Soon, an immunodiagnostic test was developed to measure the level of this protein in cancer patients. This test has been approved by the Food and Drug Administration (FDA) and used routinely to diagnose myeloma patients or to monitor the efficacy of a drug treatment in the patients by measuring the level of Bence-Jones protein. This protein decreases as the patient's condition improves in response to a particular drug treatment. Many proteins have been identified as possible markers of several diseases, however, most of them lack specificity and are not of much use in clinical medicine as diagnostics. So far, only about a few proteins are approved by the FDA as biomarkers of different human diseases; these are listed in Table 6.1.

The lack of confidence in using a particular single protein as a biomarker for a disease has led to the development of a panel of proteins as biomarkers instead of a single protein for certain diseases. It is shown that an increase in a combination of four proteins, such as leptin, prolactin, osteopontin, and insulin-like growth factor II, serves as a good indicator of ovarian cancer. None of these proteins by themselves can serve as a biomarker for the diagnosis of ovarian cancer. In addition to proteins, several hormones such as adrenocorticotropin hormone (ACTH), human chorionic gonadotropin (hCG) and calcitonin are also used as biomarkers for cancers.

Proteins have been used as drugs for many diseases or have been targets of several drugs. Insulin is the classic example of a protein used as a drug to control the level of blood sugar in persons with diabetes. Blood clotting factor and immunoglobulins are good examples of using a protein as a drug. Drugs that influence the level of a protein called "leptin" can be used to treat obesity. Gleevec, which is a specific inhibitor of kinase involved in

Table 6.1. List of proteins approved by the FDA as biomarkers of human diseases*.

Marker	Disease
1. CEA	Malignant pleural effusion
2. Her/neu	Stage IV breast cancer
3. Bladder tumor antigen	Urothelial cell carcinoma
4. Thyroglobulin	Thyroid cancer metastasis
5. Alpha fetoprotein	Hepatocellular carcinoma
6. PSA	Prostate cancer
7. CA 125	Non-small-cell cancer
8. CA 19.9	Pancreatic cancer
9. CA 15.3	Breast cancer
10. Leptin, prolactin, Osteopontin, and insulin-like growth factor II	Ovarian cancer
11. Troponin	Myocardial infarction
12. B-type natriuretic peptide	Congestive heart failure

*Based on data by Polanski and Anderson (2006).

the cell cycle, has been used effectively to treat certain cancers. Likewise, Herceptin (Genentech Inc., South San Francisco, CA) has been used to treat cancer patients.

Drug development is a branch of science that involves the use of genomics, proteomics, metabolomics, bioinformatics, structural chemistry including X-ray crystallography, synthetic chemistry, pharmacology, microbiology, biotechnology, and medical sciences. The first step in the development of a drug requires the identification of the cause of a disease, which can be readily done by genomics, proteomics, and metabolomics. Genomics can pinpoint the cause of disease by identifying the defective gene underlying a particular disease. Proteomics can pinpoint the defective protein and its nature. Metabolomics can throw light on the biochemical interaction of genes and can show how the protein products of genes have been disrupted in a particular disease. Metabolomics can also pinpoint the disruption in metabolic path ways caused by certain environmental factors, even in organisms without any genetic effects. Thus, understanding the nature of proteins and their interaction are the keys to understanding the cause of a disease and to finding a drug for the cure or amelioration of a disease. Proteins themselves are the cause of a disease, drugs for the disease, or target for the drugs used to treat the disease; proteomics is crucial to understand a disease and its treatment (Martin 1999). Bioinformatics is used in the design of drugs and in their final selection as a drug candidate, which is then used for biochemical and toxicological tests in model animal system and then in human before its approval by the FDA. Because of the complexity of the situation and the multiple lines of approach involving several branches of science, drug development is a lengthy process costing several millions of dollars over a period of several years.

6.3.3 Proteomics and Personalized Medicine

Before the advances in genomics and proteomics, most human diseases were treated based on the approach of "one size fits all." For example, today most cancer patients are subjected to a regimen of treatment involving surgery, radiation, and chemotherapy. However, with the advances in molecular biology, there is a better understanding of the metabolic basis of cancer. As a result of these developments, it has been possible to treat certain cancer patients with specific drugs like tamoxifen, Gleevec, Herceptin, aromatase inhibitors (AIs), Erbitux (ImClone Systems, Inc., New York, NY), and Vectibix (Amgen, Inc., Thousand Oaks, CA), based on the nature of cancerous formation. With the advances in genomics and proteomics, it is possible to predict the drug response of a patient. It is also possible to select a drug with the least number of undesirable side effects. It has been shown recently that certain breast cancer patients cannot benefit from a treatment with tamoxifen, as this drug is metabolized in a different way in some patients because of certain proteins made by the mutant genes. In the human body, tamoxifen is converted to endoxifen, which has the cancer-fighting property. This conversion is carried out by an enzyme encoded by the 2D6 gene. Patients with a mutation in the 2D6 gene make the converting enzyme defective and lack the ability to produce endoxifen from tamoxifen. Hence, these patients cannot benefit from the treatment of tamoxifen. Likewise gefitinib, which is an inhibitor of tyrosine kinase of the epidermal growth factor receptor (EGFR) belonging to the erb1 family, has been found to be more effective in the treatment of certain lung cancers in patients of Asian ancestry, and particularly those patients who are nonsmokers. Other findings, such as the fact that Herceptin can be prescribed to only those cancer patients who carry a mutation in the Her2 gene, confirm the role of the genetic background of a patient in devising a tailored treatment. Based on the DNA sequence of a patient, it is possible to identify patients with mutation(s) in genes that may interfere with the use of a drug and its side effects. Thus, it is possible to tailor the personalized treatment of a patient based on the DNA sequence of that individual. Thus, the advances in genomics and proteomics offer the possibility of bringing personalized medicine to the bedside of a patient.

6.3.4 Proteomics of Obesity

Obesity is one of the major human health problems in the world and that particularly in the United States. Obesity usually leads to diabetes, heart and other cardiovascular problems. It has been shown that human muscle cells or myocytes are differentiated into white fat cells that store the fat and into brown fat cells that act as a sink for sugar molecules in our body and burn fats. The brown fat cells are rich in mitochondria which possess iron that gives these cells the brownish tinge. As we grow we lose brown fat cells which leads to obesity. Recently, Bruce Spigelman and his group at the Harvard University (Spigelman et al. 2009) have identified two proteins that are responsible for the differentiation of fat cells from myocytes and from fibroblasts as well. Of these two proteins, one is identified and characterized as 140 kilodalton Zn-finger protein. Spigelman group has experimentally shown that brown fat cells are formed when this Zn-finger protein is activated in fibroblasts. They have been successful in transplanting these brown cells in obese mice. These workers have concluded that PRDM16 (Containing the Zn-finger protein) in conjunction with C/CRB-beta complex causes the formation of brown fat cells from the myoblasts and fibroblasts. Thus the proteomics seems to provide insight into cause and cure of obesity.

6.3.5 Metabolomics

Metabolomics is the study of the metabolites. Metabolites are produced as the intermediary and end products of a biochemical reaction in any metabolic pathways carried by the proteins. Genomics, transcriptomics, and proteomics can only suggest the possibilies of gene action but cannot determine the occurrence and the extent of the gene action(s). Metabolomics provides the true measure of the gene action. The role of metabolites in controlling the phenotype of an organism was first indicated by the work of Gorrod (1909) with alcaptonuria in human infants. With the study of phenylketonuria, it became obvious how intermediary metabolites, such as an excess of phenylalanine, can cause mental retardation and provided the clue for the therapeutic management or prevention of mental retardation in humans. There are about 23,000 genes and more than 100,000 proteins but only about 10,000 metabolites in humans; therefore, metabolomics seems to be an easier way to approach problems in gene action than genomics and proteomics. Metabolimics is also useful in determining the toxic effects of a substance and is of great value in toxicology. It is also useful for probing into the adverse effects of drugs and for using as a marker for the diseases. Metabolites are found in the body fluids such as urine, saliva, and plasma with the application of mass spectrometry or nuclear magnetic resonance (NMR) in conjunction with liquid chromatography. A databank of metabolites has been established for humans and plants.

6.4 METAPROTEOMICS AND HUMAN HEALTH

As a result of advances in genomics, the DNA sequences of many microorganisms that are available as pure cultures in the laboratory are known. The DNA sequences of these cultured microorganisms and the encoded proteins are assembled in the gene banks and in protein databanks. This amount, however, represents only about 1% of the total microbial species because it has not been possible to culture most of the microbial species. Thus, the only way to get genomics and proteomic information of these microorganisms is to analyze the mixed culture present as a community in a particular environment. Such genomic and proteomic analyses of a community of microorganisms are called metagenomics and meta- proteomics, respectively. The first study of a community of microorganisms was started with the analysis of 16SrRNA and later by DNA sequencing methods when the methodology for pyrosequencing became available. The metagenomics of microorganisms is advanced and has been of great aid in the taxonomy of microbial species and the assignment of a protein as soon as it is identified by proteomics to a particular microbial species. However, the metaproteomics of microorganisms has just begun to use 2D gel and spectrometric methods. The study of metaproteomics is more important than that of metagenomics. The study of metagenomics only shows the presence of a gene but the mere presence of a gene does not ensure its expression, which controls the manifestation of a phenotype. On the contrary, proteomics can provide a map of metabolic pathways in organisms at a particular time in a particular environment. Thus, to understand the manifestation of a phenotype, the study of metaproteomics is important. Moreover, metaproteomics is important not only because these microbial species cannot be obtained as pure culture but also because these different organisms work in concert to determine a phenotype. Therefore, their gene expression must be studied together.

The study of metaproteomics is important because many microorganisms exist in symbiotic relationship with the human body. Currently, metaproteomics is the only way to understand the different metabolic pathways and their interactions among themselves as well as with the biochemistry of the host, which leads to the determination of a particular phenotype. Some examples of this include human gut microbiome, human oral flora, and the urinary community of microorganisms. The relation of the microorganism community is well documented in humans. The human gut flora varies tremendously in normal and obese persons as reflected in the ratio of different constituents of the microbiome. Usually, a change occurs in the ratio of bacteriodetes to firmicutes in the range of 0.26 to 1.36; the obese persons show a low ratio bacteriodetes to firmicutes in their gut flora. This ratio can, however, be manipulated by a decrease in caloric intake by the obese individuals. The gut flora also influences the resistance to insulin in human individuals and thus may control the onset of diabetes type II. This also seems to be influenced by the changes in the choline metabolism by microflora in the human gut. In addition, the secondary metabolic products of microflora are useful in two ways: providing the dietary supplement to the host and influencing the use of several drugs as well as their interactions with certain drugs. The microflora are thus important in development of certain drugs. Different human populations and ethnic groups show a difference in the constituents of microflora, which suggests that the host genome plays a role in determining the nature of the host microflora. These differences in the nature of microflora have been correlated to the findings of proteomic analysis using 2D gel and spectrometry of human gut microflora.

II. APPLICATION OF PROTEOMICS IN BIOTECHNOLOGY, INDUSTRY, & MORE

6.5 PROTEOMICS IN BIOTECHNOLOGY AND INDUSTRY OF DRUG PRODUCTION

6.5.1 Biotechnology of Drug Production

Biotechnology has been used for the production of drugs since the 1940s. It was used to produce antibiotics. Biotechnology got another boost in drug production with the coming of recombinant DNA technology and gene cloning. Insulin was the first protein therapeutic that was produced with the use of biotechnology by the Eli Lilly Company. Soon, several protein drugs, such as human growth hormone, blood clotting factors for hemophiliacs, and erythropoietin, were produced with the use of biotechnology by companies like Genentech in California. With the advances in proteomics, the structure of any protein can be determined and used for mass production via biotechnology. During the process of drug development, currently the drug companies target the proteins for which other effective drugs are already known. The advances in proteomics are changing this approach, which is slowing the process of drug development. Proteomics is revolutionizing drug development by making several proteins available as targets for drug development. However, proteomics-based drug development biotechnology is still in its infancy.

6.5.2 Proteomics in Industry

Proteomics is playing a major role in the pharmaceutical industry/ biotechnology of drug development. This role of proteomics is based primarily on its ability to provide throughput technology for the identification of proteins as well as their modifications and interactions. Proteomics can provide information about many proteins simultaneously. This increases the number of proteins to be used as targets for the development of drugs. Its role is increased, by robotics methods, which are used in for protein identification by proteomics. In addition to the role of industrial proteomics in search of new drugs and their production, the use of proteomics in academic research also supports the industry involving the instruments of proteomics. It is estimated that more than 300 companies with annual revenue of about 10 billion dollars are involved in industrial proteomics. At times it is difficult to separate the roles of different companies because many of them not only involve proteomics but also involve genomics, transcriptomics, and above all, bioinformatics.

6.6 METAPROTEOMICS OF MICROBIAL FERMENTATION

As mentioned, symbiosis within the microbial community is useful for human health. It is known that bacterial colonies are established in the human gut within days of the infant birth and help to provide nutrition and immunity. In addition, an understanding of the metaproteomics of bacteria and fungi is important in evaluating their roles in biotechnology and industry. This study involves many kinds of products, such as various agricultural products and the development of different drugs, including antibiotics and other therapeutics. The study of microbial proteomics is also important in understanding the processes of fermentation, which controls the products of metabolites for human use. Fermentation is also responsible for biodegradation/bioremediation and the activity of biomass in energy production. Because *Escherichia coli* and many other bacteria are important in the understanding of molecular biology and in their use in several technological applications, the proteomes of *E. coli* and that of several members of Bacillaceae have been almost established fully as model systems. An understanding of the proteomes of these model bacterial systems has been useful in the study of metaproteomics of microbial communities. The study of these model bacteria has provided insight into the proteomics and synthesis of certain amino acids and other supplements that are used by humans. For example, the study of *E. coli* elucidates the biochemical steps involved in the synthesis of the amino acid threonine.

6.6.1 Metabolite- Production

The process of fermentation provides us with food, beer, wine, and drugs including several antibiotics. The efficiency of the processes involved in fermentation is bound to be increased with a better understanding of the proteomics underlying the involved biochemical steps. With this view in mind, the proteomics of wine making and that of drug production in industrial settings has been undertaken using 2D gel and spectrometry. The proteome analysis of wine-making yeast strains shows three types of enzymes under stress of glucose exhaustion. These include repressed proteins, induced proteins, and the class of proteins that undergo self-proteolysis. Most of the glycolytic proteins undergo self-proteolysis. Such an analysis has identified the roles of vacuoles and that of proteosomes in the autocatalytic degradation of enzymes under stress. These results provide insight into proteins that are required to increase the process of industrial fermentation used for making wine.

Likewise, the proteome analysis of *Bacillus subtilis* has been undertaken during bacterial growth in complex media leading to production of antibiotic gramicidin. The role of different groups of proteins has been identified in this process. The largest group of proteins includes those that are induced by glucose depletion in the medium or by alternative carbon and nitrogen sources. This includes the increase in the enzymes of the TCA (Tri Carboxylic Acid) or Krebs cycle and the decrease in the enzymes of the glycolytic pathway. This transition is initiated by amino acid starvation causing Rel-dependent repression of proteins involved in the process of translation and induction of amino acid biosynthetic pathways.

6.6.2 Bioremediation

Microbial fermentation is also used in the purification of our drinking water. The purification of water by biological wastewater treatment (BWWT) plants is the largest biotechnological undertaking on earth. This process is used to remove organic carbons, phosphorus and nitrogen-based nutrients, and several toxic substances. It requires mediation by both anaerobic and aerobic bacteria, which remove phosphorus by polyphosphate accumulation in the bacteria. There is a tremendous interest in understanding the changes in proteomes of the bacteria during the process of water purification involving enhanced biological phosphorus removal (EBPR). Many highly expressed proteins have been identified in the process of EBPR by proteomic analysis.

Bioremediation is also required for the cleanup of our environment. Bioremediation is mediated by the participation of several microbial communities. These microbial communities are required to break down several polyaromatic hydrocarbons (PAHs) and other toxic, mutagenic, and carcinogenic substances introduced into our environment by human activity. The removal of these substances from used water is critical before it is returned to our rivers and seas. The activities of microbial communities in the process of bioremediation are complex. However, a metaproteomic approach has been adopted to understand the role of the proteomes of different organisms in this process. Such an approach has identified a group of 20 proteins that are induced in *Acinetobacter/wolffi* using aniline as the sole carbon source. These proteins participated in several metabolic pathways leading to degradation of aniline, which included proteins of the beta-ketoadipate pathway, malate dehydrogenase, ATP-binding cassette (ABC) transporter, and many other proteins. Likewise, *Pseudomonas putida* when exposed to phenol showed expression of more than 80 proteins on 2D gel; of these, more than 60 proteins were increased in amount, whereas others were decreased in intensity. The group of upregulated proteins included those involved in oxidative stress, fatty acid and membrane biosynthesis, and energy and transport metabolism.

The proteome of soil microbiota has been elucidated recently using 2D gel and spectrometry at the University of California. These investigators have examined the community of soil microorganisms in a biofilm from a mine. They have identified about 1200 proteins; one important protein identified among them is a novel cytochrome that is responsible for the oxidation of iron and for the formation of the biofilm itself. The metaproteomics of microbial communities involved in the bioremediation of soil and water, particularly those in ground and estuaries, are difficult.

6.6.3 Biomass and Energy Production

Biomass is renewable in nature and is a good source of energy; it is also a viable alternative to fossil energy with a smaller carbon emission. Energy production from biomass usually involves a two-step process. In the first stage, the polymeric carbon molecules, such as cellulose, hemicellulose, starch, and lignins, are degraded into monomeric sugar molecules by the action of bacteria and/or fungi. In the next step, sugar molecules are fermented to yield alcohol, which is used as an energy source. The fuel produced from the biomass is called biofuel because of its biological origin. It is estimated that 1 ton of crushed sugarcane residue called bagasse obtained during the industrial production of sugar yields about 112 gallons of alcohol, which is then used as a biofuel. In addition to sugarcane, there are several other major sources of biofuel, which include straw, corn stalks, grains, and husks, as well as several grasses, including stretch grass. The biochemistry of the degradation of polymeric sugar molecules is not fully known except for the role of certain enzymes produced by the bacteria and fungi. Metagenomic and metaproteomic approaches are used to identify the roles of different bacteria and fungi. The proteomic approach is being applied to understand the process of fermentation. The fermentation of plant residue is complexed because xylose is one of the constituents of the monomeric sugar molecules obtained after the first step involving their biodegradation. In addition, plants contain hemicellulose and polyphenolic lignin, which produces complex production on degradation.

6.7 BEEF INDUSTRY

At the time this book was published, the beef industry was suffering a slump for many reasons including the general preference against red meat and the fact that industry has no firm way to predict the quality of meat, which depends on the tenderness, juiciness, and flavor of the beef. A concentrated effort is being made by the industry, U.S. Department of Agriculture, and the consumers to devise ways to predict the tenderness of meat. It seems that proteomics may discover a technique to predict beef tenderness. In several laboratories, the mass spectrometry approach is used to identify the proteins that are responsible for the tenderness of beef. It is known that the meat undergoes certain biochemical changes after the cow is slaughtered. Proteomic approaches are directed toward identifying these proteins that control and prevent tenderness. Recent works by McDonald Wick (Wick 2009) at the Ohio State University (OSU) has found six bands of proteins from the Angus cross steers by using electrophoresis. These bands were identified by mass spectrometry to contain about 30 proteins that control the tenderness of the meat. This work confirms previous works indicating the general roles of proteins in conferring this quality to meat, but it goes beyond in identifying the protein responsible for this feature. Results of the OSU work are expected to provide clues to predict the tenderness of meat and help the beef industry. A similar work in Norway has identified a protein that may prevent tenderness in certain other breeds of the cow.

6.8 BIOTERRORISM AND BIODEFENSE

The use of viral and microbial pathogens as weapons is prohibited by international treaties. However its feasibility cannot be ignored after the acts of terrorism on and after 9/11, as well as the anthrax attack through the mail system in the aftermath of 9/11, which led to the death of five persons in the United States. To protect against acts by terrorists, a proteomic approach has been initiated to understand the proteomics of such pathogens as well as the proteomics of survivors of pathogen attacks after receiving anthrax in the mail. Several university research departments and companies with financial support from the U.S. Department of Defense and National Institute of Allergy and Infectious Diseases (NIAID) have developed proteomic approaches. The NIAID has created a list of 160 pathogens divided into three groups. The major objectives of these studies are (a) to identify antibody against pathogens for example anthrax, (b) to identify the pathogen proteins to be used as therapeutics or targets of therapeutics, and (c) to develop the blood/serum proteomes of survivors of pathogen attacks to recognize the possible targets of anthrax and that of other pathogens, such as the plague. The investigators at Albert Einstein College of Medicine have conducted a spectrometric analysis of proteins from two pathogens *Toxoplasma gondii* and *Cryptosporidium parvum*, which interact in vivo. The pathogen proteins were crosslinked and then identified by spectrometry after protease digestion. In this way, many proteins have been identified. Among these proteins, the ones that interact with several other proteins have been identified as possible targets by drugs. Proteomics of host and pathogen and interaction of their proteomes is important to develop a biodefense strategy against acts of terrorism. There is a new dimension to the spread of germs by terrorists because the many viral pathogens and perhaps microbial pathogens because of the recent advances in synthetic biology.

Synthetic biology uses the DNA sequence information of a virus to resynthesize the viral molecule via throughput technology of nucleotide synthesis and robotics. Using these technologies, it is possible, for example, to spread the small pox virus, and cause mass chaos which has been already eliminated from the earth. Thus, the power of synthetic biology has to be protected from terrorists just like the nuclear bombs falling in their hands.

REFERENCES

Andersen J.S., Mann M. Organellar proteomics: from inventory to insight. EMBO Rep. 2006; 7: 874–879.

Anderson, W. F. 1992. Human gene therapy. Science 256, 808–813.

Arrell, D.K. I., Neverova, and J.E Van Eky. 2008. Cardiovascular proteomics Circ. Res. 88, 763–773

Bence-Jones, H. 1847. Papers on chemical pathology. Lecture III. Lancet ii, 269–272.

Goh, K, M.E. Kusick, D. Valle, B Childs, M. Vidal and A-L Barabasi. 2007. The human disease network. Proc. Natl. Acad. Sci. U.S.A. 104, 8685–90.

Gorrod, A.E. 1909. Inborn Errors of Metabolism. Oxford, UK: Oxford University Press.

Kyle, R.A. 1994. Multiple myeloma: How did it begin? Mayo. Clinic Proc. 69, 680–683.

Martin, P.J. et al. 1999. Proteomics as a major new technology for the drug discovery process. Drug Discovery Today. 4, 55–62.

McKusick, V.A. 1966. Mendelian Inheritance in Man, 12th ed. Baltimore, MD: Johns Hopkins University Press.

Polanski, M., and N. Leigh Anderson. 2006. A list of candidate cancer biomarkersfor targeted proteomics. Biomarker Insight 1, 1–48.

Spigelman, B. et al. 2009 Nature *in press*.

Wang, H. et al. (Goh et al. 2007) 2007. Characterization of mouse proteome using global proteomic analysis complemented with cysteinyl peptide enrichment. J. Proteome Res. 10, 1–13.

Wick, M. 2009. Mass spectrometry based approach to identify proteins associated with tenderness of meet in Angus cross Steers. J. Agr. Food Chem. June (in press).

Yan, W. et al. 2004. Adata set of human liver proteins identified by protein profiling via ICAT and tandem mass spectrometry. Mol. Cell. Proteom. 3.10, 1039–1041.

FURTHER READING

Barabasi, A.L. 2007. Network medicine—From obesity to diseasome. New. Engl. Med. 357, 404–407.

Brunet, S., P. Thibault, E. Gagnon, P. Kearney, J.J. Bergeron, and M. Desjardins. 2003. Organelle proteomics: Looking at less to see more. Trends Cell Biol. 13, 629–638.

Cho, W.C.S. 2007. Contribution of oncoproteomics to cancer biomarker discovery. Mol. Cancer, 25, 1476–1489.

Datta, S., D. Turner, R. Singh, L.B. Ruest, W.M. Pierce, and T.B. Knudson. 2008. Fetal alcohol syndrome(FAS) in C57BL/6mice detected through proteomics screening of the amniotic fluid. Clin. Molec. Teratol., 82, 177–186.

Daub, H. K. Godl, D. Brehmer, B. Kelb, and G Muller. 2004 Evaluation of Kinase inhibitor selectivity by chemical proteomics. Assay Drug DEV. Technol. 2, 215–224.

Figeys, D. 2008. Industrial Proteomics. New York: Wiley.

Godi, K., J. Wissing, A. Kurtenbach, P. Habenberger, S. Blencke, H. Gutbrod, K. Salassidis, M. Stein-Gerlach, M. Cotton, and H. Daub. 2003. An efficient method to identify the cellular targets of protein kinase inhibitors. Proc. Nat. Acad. Sci. 100, 15434–15439.

Grant, S.G.N. and W.P. Blackstock. 2001. Proteomics in neurosciences: From protein to network. J. Neurosci. 21, 8315–8318.

Hanash, S. 2003. Disease proteomics. Nature 422, 226–232.

He, F. 2002. Human liver proteome project. Mol Cell Proteom. 4, 1841–1848.

Hollung, K. 2009, Proteomics as Research Tools in Meat Science. Reading, UK: IFIS Publishing.

Huber, L.A., K. Pfaller, and I. Viector. 2003. Organelle proteomics. Circ. Res. 92, 962–968.

Hutchinson, C.A. 2007. DNA sequencing: Bench to bedside and beyond. Nuclic Acids Res. 35, 6247–6237.

Jones, H.B. 1848. On a new substance occurring in the urine of a patient with mollities ossium. Phil. Trans. R. Soc. Lond. 138, 55–62.

Kim, Y.-H., J.-S. Park, J.Y. Cho, Y.-H. Park, and J. Lee. 2004. Proteomic response analysis of a threonin-overproducing mutant of E. coli. Biochem. J. 381, 823–829.

Kwang-II, G., M.E. Cusick, D. Valle, B. Childs, M. Vidaland, and A.L. Barabasi. 2007. The human disease network. Proc. Nat. Acad. Sci 104, 8685–8690.

Klaassens, E.S., W.M. de Vos, and E.E. Vaughan. 2007. Metaproteomics approach to study the functionality of the microbiota in the human infant gastrointestinal tract. Applied Environ. Microbiol. 73, 1388–1392.

Li, M. et al. 2008. Symbiotic gut microbes modulate human metabolic phenotypes. Proc. Nat. Acad. Sci. 105, 2117–2122.

Lopez, M.F., and S. Melov. 2002. Applied proteomics: Mitochondrial proteins and effect on functions. Circ. Res. 90, 380–389.

Marko-Varga, G., et al. 2007. Personalized medicine and proteomics: Lessons from non-small cell lung cancer. J. Proteome Res. 6, 2925–2935.

Molloy, S. 2005. Malaria Secretome uncovered. Nature Re. Microbiol. 3, 97–98.

Ohlmeier, S., A.J. Kastanoitis, J.K. Hiltunen, and U. Bergman. 2004. The yeast mitochondrial proteome, a study of fermentative and respiratory growth. J. Biol. Chem. 279, 3956–3979.

Pritzker, K.P.H. 2002. Cancer biomarkers: Easier said than done. Clin. Chem. 48, 1147–1150.

Raj, U., et al. 2007. Genomics and proteomics of lung diseases. Am. J. Physiol. Lung Cell Mol. Physiol. 293, 145–151.

Reo, N.V. 2002. NMR-based metabolomics. Drug Chem. Toxicol. 25, 357–382.

Rezaul, K., L. Wu, V. Mayya, S.-II Wang, and D. Han. 2005. A systematic characterization of mitochondrial proteome from human T leukemia cells. Mol. Cell Proteom. 4, 169.

Schmidt, A. and R. Aebersold. 2006. High accuracy proteome maps of human body fluids. Genome Biol. 7, 242–246.

Singh, O.V. and N. Nagaraj. 2006. Transcriptomics, proteomics and interactomics: Unique approaches to track the insights of bioremediation. Brief. Functional Genom. Proteom. 4, 355–362.

Shi, R., C. Kumar, A. Zaugman, Y. Zhang, A. PodteleJnikov, J. Cox, J.E. Wikniewski, and M. Mann. 2007. Analysis of mouse liver proteome using advanced mass spectrometry. J. Proteome Res. 6, 2963–2972.

Tarun AS, et al. A combined transcriptome and proteome survey of malaria parasite liver stages. Proc. Natl. Acad. Sci. U.S.A (2008) 105: 305–310.

Veenastra, T.D., T.P. Conrads, B.L. Hood, A.M. Avellino, R.G. Ellenbogen, and R.S. Morrison. 2005. Biomarkers: Mining the biofluid proteome. Mol. Cell. Proteom. 4.4, 409–418.

Wilms, P., M. Wexler, and P.L. Bond. 2008. Metaproteomics provides functional insight into activated sludge waste water treatment. Plos One 3, 1–16.

CHAPTER 7

PROTEOMICS – FUTURE DEVELOPMENTS

Proteomics started with the development of methods for the purification of proteins. Initially, it involved the purification of one protein at a time mostly by column chromatography. Electrofocusing of proteins was a major development in protein purification, which led into development two-dimensional (2D) gel. Two-dimensional gel resulted in the purification of many proteins on one gel for subsequent characterization. The development of mass spectrometry, genomics, and bioinformatics brought a revolution in proteomics. Now, several proteins can be analyzed simultaneously.

7.1 TECHNICAL SCOPE OF PROTEOMICS – BEYOND PROTEIN IDENTIFICATION

Current proteomic methods of protein separation by 2D gel and/or high-performance liquid chromatography (HPLC), as well as the identification of proteins by mass spectrometry (MS) in conjunction with protein databanks and several software programs are satisfactory. These methods must be improved in the future to be more efficient. Some improvements that are expected in the near future must include the following strategies as discussed by others as well:

1. Preferential removal of high-abundance proteins to enrich a low-abundance protein.

Introduction to Proteomics: Principles and Applications, By Nawin C. Mishra

Currently, low-abundance proteins are enriched after the removal of several high-abundance proteins by immunoprecipitation. During this process, many growth factors and peptides, such as cytokines, are also removed because of the nonspecific binding of these factors to high-abundance proteins. It seems that low-and high-abundance proteins exist with a difference of 10-order magnitude. Current protein separation techniques separate proteins of 3–4 order of difference between the high- and low-abundance proteins in the human plasma proteome. Among these techniques is many low-abundance proteins that have not yet been discovered. Some of these growth factors and peptides may be potential biomarkers, and therefore, methods must be developed in the future to obtain them or separate them without their removal with the high-abundance proteins from blood.

2. Miniaturization of the platform.
   Miniaturization of the platform is required to scale down the process of protein analysis, which involves handling a much smaller amount of the sample. It may soon be possible to scale down the size of the sample from a nanoliter to a picoliter, which reduces the amount of protein from fetomoles to attomoles. Miniaturization will require use of microfluidics to handle a miniscule of sample without any loss by absorption or by evaporation. In this way, it will be possible to analyze the protein sample on a protein chip.
3. Automation.
   Progress is being made to automate the process of protein analysis by using robots to conduct programmable isoelectrofocusing (IEF) and 2D gel unit analysis and execute post handling of samples after the separation of proteins. Automation will also involve the excision of protein spots from the gel, their digestion by trypsin, and subsequent cleanup and direct placing of the samples on the MALDI-TOF-MS port.
4. New software programs and faster computers.
   It is required to develop new software and faster computers to allow more accurate detection, quantification, and identification of proteins in the gel spot.
   Methods will be developed to increase the sensitivity of current instruments to avoid falses positive or false-negative results among MS data. Currently, up to 20% of MS data included such false positives/negatives.
5. Increased sensitivity.

As mentioned in chapter 4 new software must be developed to allow more accurate detection, quantification, and identification of protein in the gel spot.

6. Artificial intelligence.
   It would be possible to develop and apply artificial intelligence to identify and characterize proteins. Recently, it has been possible to apply artificial intelligence to identify many biomarkers for ovarian cancer.
7. Development of in situ mass spectrometry.
   Methods are being developed to analyze the proteins in slices of a tissue. Using this procedure, it is possible to compare the protein contents of a normal cell and that of a cell from a person with a disease. This will help to identify the protein biomarker(s) of the disease for diagnostics and for the development of a drug or to follow the effect of a drug during treatment.
8. Computer analysis of biochemical pathways.
   Most diseases involve interactions of metabolic pathways. In the future, it is possible to develop computer programs to identify the fallouts in metabolic interactions that can help in the development of diagnostic as well as the drug.

7.2 SCIENTIFIC SCOPE OF PROTEOMICS – CONTROL OF EPIGENESIS

Epigenomics includes changes in the genomes by proteins as well as by small molecules. Understanding epigenomics holds the key to answering important questions in biology. This is particularly so in the area of differentiation and development. It may explain how myriads of cell types in the multicellular organisms possessing the exact same genome differ in their structure and function. In addition, epigenetic changes are responsible for cancer and other human diseases, as well as the process of aging in humans. Proteins provide a distinct role in the understanding of these questions because of their role in the nucleosomal organization of the chromosomes. The nucleosomal organization of chromosomes transcribes certain genes while masking other genes, which leads to the differentiation of cell types and their organization into different tissues. In addition, proteins working as transcription factors add another dimension to the differentiation of cell types. Shinya Yamanaka and his collaborators (Takahashi et al. 2007) in Japan have converted normal skin cells in to pluripotent stem cells by introduction of four genes for transcription factors into skin cells. These stem cells, thus obtained by Shinya Yamanaka and his coworkers, were then differentiated to yield an array of tissues. These results provided an alternative to obtaining stem cells from human embryos, which a solution to offers the issues surrounding the use of embryonic stem cells by certain groups in the United States. From this study, it has become obvious that different classes of transcription factors exist. The transcription factors introduced into skin cells by this group of Japanese authors belong to a class of master transcription factors. The results of these and other studies clearly point out the role of proteomics in understanding epigenesis. Epigenesis is becoming important both from the scientific and application points of view: Many diagnostics are being developed based on the understanding of the epigenetic changes of genes. It is expected that proteomics will provide clues for the differentiation of cell types from the stem cells as well as their use in different types of therapies for different human diseases in the future. Therefore, soon, proteomics is expected to provide a full understanding of the problems of differentiation during human development in addition to its role in the creation and use of stem cells.

7.3 MEDICAL SCOPE OF PROTEOMICS

Proteomic technology has provided a deeper insight into the structure and function of proteins, including the different modifications of proteins, their interactions, and their roles in metabolic pathways. It is expected to provide subsequent insight into the causes of various diseases and their possible diagnosis, treatment, and cure. Some of these elements are discussed in this chapter.

7.3.1 Human Diseases

Proteomics hold great promises for the diagnosis and treatment of several numbers of diseases, which include cancer, cardiovascular, neurodegenerative, metabolic, and infectious diseases. Proteomics has made it possible to generate a panel of biomarkers for the diagnosis of cancer. Most of these panels include many proteins, sometimes more than 25 proteins. In the future, it may be possible to identify a smaller number of proteins as essential biomarkers. It would be possible to select the key markers by using artificial intelligence and better software applications. It would be possible to detect cancer at an early benign stage, which would make the treatment much more manageable. It would be possible to develop biomarkers that characterize slow-growing cancer from aggressive and rapidly growing cancer. Some of these markers would be identified among the proteins of signaling systems, membrane proteins, and regulators of metabolic pathways and switches. Likewise, markers for the early detection of cardiovascular, neurodegenerative, and metabolic diseases will be identified and used to detect and treat diseases. Advancement is expected in the early detection of infectious diseases by proteomic analysis, as is being developed for malaria by proteomics of the parasite. Understanding the host proteomics after an attack by an infectious agent should provide information about the management of antibiotic-resistant bacteria. Understanding the genomics and proteomics of viruses would provide a clue to treat or prevent such viral infection, for example, severe acute respiratory syndrome (SARS) and cold viruses. The genomics of more than 97 cold viruses have made it possible to identify a common region in all these viruses, which will be targeted by a drug in the future to provide a better remedy to the common cold.

7.3.2 Development of Drugs

The future of proteomics lies in the ability to design better drugs for human diseases. It is believed that proteomics will provide a better means to diagnose, treat, and monitor the efficacy of a drug treatment. It is expected that the field of proteomics will design better diagnostics and smart drugs by using the biomarkers revealed from the proteomic analysis of a normal and diseased person. The identification of protein biomarkers, their modification, and altered metabolic pathways by comparison of the proteomes of normal cell and cell from a diseased person will be used to design smart drugs. Drug development is an expensive enterprise. Proteomics is expected to reduce the cost by increasing the number of target proteins used for the drug designed. Knowledge of metabolic pathways and that of proteins interactions will be used to facilitate the development of drugs in a cost-effective manner with the help of new bioinformatics tools. In the future, the interaction of possible chemicals as drugs with the target proteins will be accessed rapidly by high throughput screening (HTS) methods in a cost-effective manner. This technique will be aided by the use of combinatorial chemistry and the library of chemicals available on the database.

Proteomics will also lead to the discovery of new proteins as drugs. This possibility is suggested by the current studies of Mark Tuszynski and his coworkers at the University of California, San Diego. The researchers have shown that the symptoms of Alzheimer disease, such as memory loss, brain cell degeneration, and cognitive impairment, can be overcome by the injection of a brain-derived neurotrophic factor (BDNF) into mice, rhesus monkeys, and other model animals. The BDNF protein is usually produced in the brain of normal animals, but it is not produced by the cortex in animals with the disease. In addition to the development of drugs, proteomics is expected to help in the production of vaccines in the future.

7.3.3 Personalized Medicine

It is hoped that proteomics will make personalized medicine a routine part of medical practice and patient care. With the advances in genomics and proteomics, it is possible to develop diagnostic strategies to predict the suitability of a drug to be prescribed to a patient for maximum efficacy. It seems at least 200 prescribed drugs exist that do not work for a particular patient or have undesirable side effects. In such cases, the patient carries a gene mutation that negates the efficacy of a particular drug. For example, the antiretroviral drug abacavir does not work in 20% of patients who carry a mutation in the HLA gene. A diagnostic test to detect the HLA variant in a population is now available, which makes it possible to prescribe abacavir only to patients who lack HLA mutation and benefit from the use of this drug. Likewise, not all persons can benefit from wafarin, which is an anticoagulating drug used to treat patients who suffer from thrombosis and other blood disorders. The dose of wafarin must be adjusted to avoid internal bleeding in such patients. A diagnostic test is available for wafarin but is not yet approved by the U.S. Food and Drug Administration (FDA). It is known that certain cancer patients cannot benefit from tamoxifen because they carry a mutation 2D6 in the gene, which produces an enzyme that converts tamoxifen into endoxifen, which is a biologically active substance with cancer-fighting effects. These breast cancer patients have to be put on another class of drugs called aromatase inhibitors. However, the use of aromatase inhibitors is limited because it works only in menopausal women. This leaves the woman with breast cancer who has not yet completed menopause with very odd choices. Other drugs must be developed for these patients. Likewise, it has been shown that certain colon cancer patients who carry a mutation called KRAS cannot benefit from the prescription of Vectibix (Amgen, Inc., Thousand Oaks, CA). Now, Vectibix is prescribed only to patients who lack the KRAS mutation. In the future, it would be routine to conduct personal genomics of an individual or require a particular diagnostic examination before a drug is prescribed. The delivery of personalized medicine will require the development of an electronic storage of the entire history of medical records of an individual as well as his/her interactions to a drug based on several diagnostics, subgroupings of disease, and genomics. The practice of personalized medicine raises many questions in medical ethics regarding the development of diagnostics for the use of drugs and the questions of cost effectiveness of developing such diagnostics. This also raises ethical questions on the part of physicians in prescribing certain treatment regimens.

7.3.4 Proteomics and Metabolomics

Metabolimics is the study of metabolites. Metabolites are the substances produced by the enzymatic action of a protein in any metabolic pathway. It is known that genes produce enzymes, and enzymes catalyze a biochemical to produce metabolites, which determine the phenotype of an organism. Thus, genomics, proteomics, and metabolomics are related. Metabolites can be used to indicate a disease condition and, thus, are helpful in developing diagnostics for human diseases. It has been found that the presence of sarcosine in human urine is an indicator of an aggressive form of prostate cancer. Human prostate cancers are usually of two kinds; one is slow growing and another is fast growing. The fast-growing kind is aggressive. Thus, the detection of sarcosine in human urine can be used to identify the aggressive form of prostate cancer at an early stage, and then it can be treated accordingly. The proteome changes in the cancer patients can be deduced from the presence of sarcosine in urine. The presence of sarcosine indicates a defect in the enzyme that converts glycine into sarcosine or changes in the protein(s) that catalyzes the degradation of sarcosine. Other effects of small molecules or metabolites on proteomics are illustrated by a recent finding at the University of California, Los Angeles (UCLA). The UCLA scientists have shown changes in the proteomics of lung cells, which affect the proteins involved in cell motility by green tea extract. These workers interpret that polyphenols in green tea extract bring changes in the proteomics of the lung cell and are responsible for inhibiting the metastasis and growth of cancer cells. Likewise, it has been shown that alcohol consumption by adolescents causes proteomic changes in the hippocampus region of the brain. These studies provide the biochemical basis for alternative medicine and the relations among metabolomics, proteomics, and human health.

The approach of metabolomics is simple, noninvasive, and much more inexpensive than the genomic and proteomic approach to detect diseases. Scientists in Israel have developed computer programs to monitor several metabolites by comparing the metabolic pathways in both normal and diseased cells. A library of metabolites is now available in the form of a databank. It is obvious that much of the proteomics will be detected by the study of metabolomics.

7.4 PROTEOMICS, ENERGY PRODUCTION, AND BIOREMEDIATION

All forms of energy on this planet are essentially bioenergy, which is energy from the sun trapped by biological means. Thus, biology, like engineering and chemistry, is important in finding solutions to our energy problems. Currently, biology is involved only in the production of biofuels, purification of water, and other environmental problems by microbial community. It is expected that in the future, advances in synthetic genomics and proteomics can solve our problems in energy production, bioremediation of environment, carbon sequestration, and other related problems. Synthetic genomics attempt to create new microbes with selected genomes suitable for energy production and environmental bioremediation. Synthetic genomics has just started: Soon, the proteomics of these newly created organisms will usher in the era of synthetic proteomics.

7.5 PROTEOMICS AND BIODEFENSE

The science of biodefense has become a reality after the attacks on the World Trade Center on September 11, 2001. It is expected that many diagnostics and vaccines against several microbes will be developed largely because of the understanding of the proteomics conducted by the U.S. Departments of Defense and of Homeland Security.

REFERENCE

Takahashi, K.K., Tanabe, M. Ohnuki, M. Narita, T. Ichisaka, K. Tomoda, and S. Yamanaka. 2007. Induction of pluripotent stem cells from adult human fibroblasts by defined factors. Cell 11, 1–12.

FURTHER READING

Anderson, N. L. 2005. The role of multiple proteomic plateforms in a pipeline for new diagnostics. Molec. Cell. Proteom., 4.10, 1441–1445.

Geho, D. H., J. N. Cooper, V. Epina, E. Garaci, and E. F Patricoin. 2006. Role of proteomics in personalized medicine. Personalized Med. 3, 223–226.

Jain, K. K. 2004. Proteomics and drug discovery. In T. V. Klein (ed.). Proteomics in Nephrology, Basel, Switzerland: Basel Karger.

Khan, J. 2003. Genomic and proteomic technological advances in cancer research. Pharmacogenomics 4, 245–249.

Knauf, M. and M. Moniruzzaman. 2004. Lignocellulosic processing: A perspective Intl. Sugar J. 106, 147–150.

Nagahara, A. B. et al. 2009. Neuroprtective effects of brain-derived neurotropic factor in rodent and primate models of Alzheimer's disease. Nat. Med. 2, 1–8.

Verrils, N. M. 2006. Clinical proteomics: Present and future prospects. Clin. Biochem. Rev. 27, 99–116.

INDEX

Introduction to Proteomics: Principles and Applications, By Nawin C. Mishra

Printed in the United States
By Bookmasters